Designing Steel Structures for Fire Safety

Designing Steel Structures for Fire Safety

Jean-Marc Franssen

Department of Architecture, Geology, Environment & Constructions, University of Liège, Liege, Belgium

Venkatesh Kodur

Department of Civil & Environmental Engineering, Michigan State University, East Lansing, USA

Raul Zaharia

Department of Steel Structures and Structural Mechanics, "Politehnica" University of Timisoara, Timisoara, Romania

CRC Press
Taylor & Francis Group
Boca Raton   London   New York   Leiden

CRC Press is an imprint of the
Taylor & Francis Group, an **informa** business

A BALKEMA BOOK

Cover photo: Courtesy of CTICM, GSE and INERIS

*Taylor & Francis is an imprint of the Taylor & Francis Group,
an informa business*

©2009 Taylor & Francis Group, London, UK

Typeset by Macmillan Publishing Solutions, Chennai, India
Printed and bound in Great Britain by TJ International Ltd,
Padstow, Cornwall

Published by: CRC Press/Balkema
                P.O. Box 447, 2300 AK Leiden, The Netherlands
                e-mail: Pub.NL@taylorandfrancis.com
                www.crcpress.com – www.taylorandfrancis.co.uk – www.balkema.nl

British Library Cataloguing in Publication Data
A catalogue record for this book is available from the British Library

Library of Congress Cataloging-in-Publication Data

Franssen, Jean-Marc.
  Designing steel structures for fire safety / Jean-Marc Franssen, Venkatesh Kodur, Raul Zaharia.
    p. cm.
  Includes bibliographical references.
  ISBN 978-0-415-54828-1 (hardcover : alk. paper) – ISBN 978-0-203-87549-0 (e-book)
1. Building, Fireproof.   2. Building, Iron and Steel.   I. Zaharia, Raul.   II. Kodur, Venkatesh.   III. Title.

  TH1088.56.F73 2009
  693.8′2–dc22

                                                                2009006378

ISBN 978-0-415-54828-1(Hbk)
ISBN 978-0-203-87549-0(eBook)

Table of Contents

Foreword

This book is a major new contribution to the wider understanding of structural behaviour in fires. The art and science of designing structures for fire safety has grown dramatically in recent years, accompanied by the development of sophisticated codes of practice such as the Eurocodes. The Eurocode documents have evolved over several decades and now represent the best international consensus on design rules for structures exposed to fires.

Codes alone do not provide enough information for designers, especially as they become more sophisticated and comprehensive. Most codes have been written by a small army of dedicated experts, some of whom have been immersed in the project for many years with the responsibility to provide the correct rules, not always providing adequate guidance for using those rules. Designers want to understand the basic concepts of the code and the philosophy of the code-writers, together with hands-on advice for using the code.

Structural design for fire is conceptually similar to structural design for normal temperature conditions, but often more difficult because of internal forces induced by thermal expansion, strength reduction due to elevated temperatures, much larger deflections, and many other factors. Before making any design it is essential to establish clear objectives, and determine the severity of the design fire. This book shows how these factors are taken into account, and gives guidance for all those wishing to use the Eurocodes for fire design of steel structures.

Prof. Jean-Marc Franssen has been a pioneer in the field of structural design for fire safety, with extensive involvement in codes, software development, research, teaching, and consulting. He is also well known for establishing the SiF Structures in Fire international workshops. His co-authors are Dr Raul Zaharia, a leading European researcher in structural fire engineering, Prof. Venkatesh Kodur from Michigan State University who is one of the top researchers and teachers in structural fire design in North America. Together they have produced a book which will be extremely valuable to any design professionals or students wishing to use and understand Eurocode 3, or to learn more about the design of steel structures exposed to fires.

The fire sections of the Eurocodes are considered to be among the most advanced international codes of practice on fire design of structures, and have attracted attention around the world. This book is an excellent introduction for readers from other regions who wish to become knowledgeable about the philosophy, culture and details of the Eurocodes for structural fire design.

Professor *Andy Buchanan*
University of Canterbury, New Zealand

Preface

Fire represents one of the most severe conditions encountered during the life-time of a structure and therefore, the provision of appropriate fire safety measures for structural members is a major safety requirement in building design. The basis for this requirement can be attributed to the fact that when other measures for containing the fire fail, structural integrity is the last line of defence.

The historical approach for evaluating fire resistance of structural members is through prescriptive-based methodologies. These methodologies have significant drawbacks and do not provide rational fire designs. Therefore, in the last two decades there has been important research endeavours devoted to developing better understanding of structural behaviour under fire conditions and also to develop rational design approaches for evaluating fire resistance of structures. This activity was particularly significant in Western Europe where numerous research reports, Ph.D. theses and scientific papers were published.

European technical committees were in the fore-front to implement some of the research findings in to codes and standards to enable the application of rational fire engineering principles in the design of structures. Among the first internationally recognised codes of practice are, for steel elements, the recommendations of the ECCS "*European Convention for Constructional Steelwork*" (ECCS 1983) and, for concrete elements, the recommendations of the CEB/FIP "*Comité Euro-International du béton / Fédération Internationale de la précontrainte*" (CEB 1991). The fire parts of the Eurocodes were first presented in Luxemburg in 1990. Over the next few years these Eurocode documents have been significantly updated by incorporating new or updated provisions based on latest research findings reported from around the world.

On similar lines, in the last few years, many countries have moved towards implementing rational fire design methodologies in codes and standards. One such example is the recent introduction of rational fire design approach in the latest edition of American Institute of Steel Construction's steel design manual. In addition, a number of countries around the world are updating their codes and standards by introducing performance-based fire safety design provisions. A performance-based approach to fire safety often facilitates innovative, cost-effective and rational designs. However, undertaking performance-based fire safety design requires the advanced models, calculation methodologies, design manuals books and trained personnel.

The Eurocode documents, or recently updated codes and standards in other countries, are nevertheless far from being useful textbooks, lecture notes or guidance documents. While these codes and standards provide specifications for undertaking rational fire design, there is no detailed commentary or explanation for the various specifications or calculation methodologies. Added to this, fire design is rarely taught as part of regular engineering curriculum and thus most engineers, architects and regulators may not be fully versed with the necessary background to easily understand the relevant clauses, or to make interpretations, or to recognise the limits of application of various rules. In other words, unless one has some level of expertise in fire safety engineering, it is not easy to apply the current provisions in codes and standards in most practical situations. Compounding this problem is the fact that there is only handful number of text books in the area of structural fire engineering.

This book is aimed at filling the current gaps in structural fire engineering by providing necessary background information for rational fire design of steel structures. It deals with various calculation methodologies for fire design and analyses structural steel elements, assemblies and systems. The intent is to provide a basis for engineers with traditional backgrounds to evaluate the fire response of steel structures at any level of complexity. Since the main aim of the book is to help facilitate rational fire design of steel structures, the book relies heavily on Eurocode 3 provisions, as well as relevant fire provisions in American and other codes and standards.

In this book the information relevant to fire design of steel structures is presented in a systematic way in seven chapters. Each Chapter begins with an introduction of various concepts to be covered and follows with detailed explanation of the concepts. The calculation methods as relevant to code provisions (in Europe, North America or other continents) are discussed in detail. Worked examples relevant to calculation methodologies on simple structural elements are presented. For the case of complete structures guidance on how analysis can be carried out is presented.

Chapter 1 of the book is devoted to providing relevant background information to codes and standards and principles of fire resistance design. The chapter discusses the fire safety design philosophies, prescriptive and performance-based design fire safety design issues. Chapter 2 deals with basis of design and mechanical loads. The load combinations to be considered for fire design of structures, as per European and North American codes and standards, are discussed.

Chapter 3 discusses the detailed steps involved in establishing the fire scenarios for various cases. Both Eurocode and North American temperature-time relationships are discussed. Procedures in this section allow the designer to establish the time-temperature relationships or heat flux evolutions under a specified design fire. Chapter 4 deals with steps associated in establishing the temperature history in the steel structure, resulting from fire temperature. The various approaches for undertaking thermal analysis by simple calculation models are discussed.

Chapter 5 presents the steps associated for establishing the mechanical response of a structure exposed to fire. The possibilities for analysis at different levels: member level, sub-structure level and global level are discussed. Full details related to simple calculation methods for undertaking strength analysis at member level are presented. Chapter 6 is devoted to fire resistance issues associated with design of joints. The steps associated with the fire resistance of a bolted or a welded joint through simplified and detailed procedure are discussed.

Chapter 7 deals with thermal and mechanical analysis through advanced calculation models. The procedures involved in the sub-structure analysis or global structural analysis under fire exposure is fully discussed. Case studies are presented to illustrate the detailed fire resistance analysis of various structures. Chapter 8 presents four design examples showing how a complex structure can be designed using the concept of element or sub-structure analysis.

The book concludes with two Annexes which present some of the design information related to material properties and temperature profiles. Annex 1 focuses on thermal properties of structural steel and commonly used insulation materials and resulting temperature profiles in steel. Annex 2 focuses on mechanical properties of structural steel.

This text book is a reference that allows designers to go beyond current prescriptive approaches that generally do not yield a useful understanding of actual performance during a fire, into analyses that give realistic evaluation of structural fire performance. The book is a compendium of essential information for determination of the effects of fire on steel structures. However, the book is not a substitute for the complete text of Eurocode 3 or any other codes and standards. The book should help a reader not familiar with fire safety engineering to make relevant calculations for establishing the fire response of steel structures.

The target audience for this book is professionals in engineering or architecture, students or teachers in these disciplines, and building officials and regulators. A good knowledge of mechanics of structures is essential when reading this book, while general background on the design philosophy related to building structures is an advantage. It is hoped that the book will enable researchers, practioners and students to develop greater insight of structural fire engineering, so that safer structures could be designed for fire conditions.

If needed, an errata list will be placed on *www.structuresinfire.com*.

Jean-Marc Franssen, Venkatesh Kodur, Raul Zaharia
jm.franssen@ulg.ac.be
kodur@egr.msu.edu
raul.zaharia@ct.upt.ro

Notations

Latin upper case letters

A	cross-sectional area of a member
A_m	surface area of a member per unit length
A_p	appropriate area of fire protection material per unit length of the member
A_t	area of walls, ceiling and floor including openings
A_v	total area of vertical openings of all walls
D	characteristic length of the fire
$E_{a,\theta}$	Young's modulus
E_d	design value of effects of actions at room temperature
$E_{fi,d}$	design value of effects of actions in case of fire
$F_{d,fi}$	design value of the actions in case of fire
$F_{b,Rd}$	design bearing resistance of a bolt at room temperature
F_{ij}	view factor
$F_{t,Rd}$	design tension resistance of a bolt at room temperature
$F_{v,Rd}$	design shear resistance of a bolt per shear plane at normal temperature
$F_{w,Rd}$	design strength of a fillet weld at normal temperature
$G_{d,fi}$	design values of the permanent actions in case of fire
G_k	characteristic value of the permanent action
H	vertical distance between the fire source and the ceiling
I	second moment of area of a cross-section
L_f	flame length
L_h	horizontal flame length
O	opening factor
$P_{d,fi}$	design values of the prestressing action in case of fire
P_k	characteristic value of the prestressing action
Q	rate of heat release of the fire
Q_c	convective part of the rate of heat release
Q_D^*	non-dimensional square root of the Froude number
Q_H^*	non-dimensional square root of the Froude number
$Q_{d,fi}$	design values of the variable actions in case of fire
Q_k	characteristic value of the variable action
$R_{fi,d,t}$	design value of the resistance in case of fire
RHR_f	rate of heat release densities

V	volume of a member per unit length
V_{RD}	shear resistance of the gross cross-section for normal temperature design
W_{el}	elastic modulus of a section
W_{pl}	plastic modulus of a section
$X_{d,fi}$	design value of the material properties in case of fire

Latin lower case letters

b	wall factor
c	specific heat
d_p	thickness of the fire protection material
$f_{p,\theta}$	limit of proportionality
f_y	yield strength
$f_{y,\theta}$	effective yield strength
$\dot{h}$	heat flux received by the structure at the level of the ceiling
$\dot{h}_{net}$	net heat flux
h_{eq}	weighted average of opening heights on all walls
$k_{b,\theta}$	reduction factor for bolts
$k_{E,\theta}$	reduction factor for the Young's modulus
k_p	parameter for protection material, see Eq. 4.9
$k_{p,\theta}$	reduction factor for the limit of proportionality
k_{sh}	correction factor for the shadow effect
$k_{w,\theta}$	reduction factor for weld
$k_{y,\theta}$	reduction factor for the effective yield strength
l_{fl}	buckling length
$q_{t,d}$	design value of the fire load density
r	horizontal distance from the vertical axis of the fire to the point under the ceiling where the flux is calculated
t	time
t^*	modified time in the parametric fire model
t_{lim}	shortest possible duration of the heating phase
t_{max}	duration of the heating phase
y	ratio between two distances
z'	vertical position of the virtual source
z_0	position of the virtual origin

Greek upper case letters

β_M	equivalent uniform moment factor
Γ	factor in the parametric fire model
χ	buckling coefficient
χ_{LT}	coefficient for lateral torsional buckling
ψ_0	coefficient for combination value of a variable action, taking into account the reduced probability of simultaneous occurrences of the most unfavourable values of several independent actions

ψ_1 coefficient for frequent value of a variable action, generally representing the value which is exceeded with a frequency of 0.05, or 300 times per year

ψ_2 coefficient for quasi-permanent value of a variable action, generally representing the value which is exceeded with a frequency of 0.50, or the average value over a period of time

Greek lower case letters

α imperfection factor, see Eq. 5.14
α_c coefficient of heat transfer by convection
γ_G partial factors for the permanent actions
γ_P partial factors for the prestressing action
γ_Q partial factors for the variable actions
ε factor for local buckling, see Eq. 5.5 and 5.7
ε_m surface emissivity of a member
κ_1 adaptation factor for non-uniform temperature in a cross-section
κ_2 adaptation factor for non-uniform temperature along a beam
η_{fi} conversion factor between the effects of action at room temperature and the effect of action in case of fire
θ_g gas temperature in the compartment or near a steel member
$\theta_{m,t}$ surface temperature of a member at time t
λ thermal conductivity
$\overline{\lambda}$ non-dimensional slenderness
ρ density
μ_0 degree of utilisation in the fire situation
σ Stephan Boltzmann constant

subscripts
0   independent action, virtual origin
1   frequent action
2   quasi-permanent action
A   accidental
a steel
b box
c convection
d design
el elastic
f flame
fi fire
G   permanent
h horizontal
g gas
k characteristic
lim limit
m member

max	maximum
P	prestressing
p	protection material, proportionality
pl	plastic
req	required
sh	shadow
t	total
v	vertical

Author profiles

Jean-Marc Franssen is professor at the University of Liege in Belgium from where he graduated in 1982. He has specialised his research career on the behaviour of structures subjected to fire which was the subject of his Ph.D. thesis (Franssen, 1987), and his *"thèse d'agrégation de l'enseignement supérieur"* (Franssen, 1997). During the last 20 years, he has been involved in numerous research programs on the fire behaviour of steel and composite steel-concrete structures, a lot of them with the support of the ECSC *"European Coal and Steel Community"*. The steel industry, namely the Luxemburg company ARBED, now in the ARCELOR-Mittal group, played a prominent role in several of these projects, with CTICM from France, TNO from the Nederland and LABEIN from Spain as regular partners. At the University of Liege, he holds the chair of Fire Safety Engineering. He was a member of the Project Team for transforming the fire section of Eurocode 3 from an ENV to an EN, he played a key role in writing the Belgium National Application Documents to the fire parts of the Eurocodes on steel, concrete and composite structures, and has been an active member of two Technical Group chaired by Prof. U. Schneider within RILEM, *"Réunion International des Laboratoires d'Essais sur les Matériaux"*, namely *"TC 129-MHT: Test Methods for Mechanical Properties of Concrete at High Temperatures"* and *"TC HTC Mechanical Concrete Properties at High Temperature – Modelling and Applications"*. He initiated the international *"Structures in Fire"* workshops held in Copenhagen, Christchurch, Ottawa, Aveiro and Singapore from 2000 to 2008. Having been involved in the *"Natural Fire Safety Concept"* series of research projects, his expertise goes beyond structural behaviour to include modelling of fire growth and severity. He has acted as the Belgian National Technical Contact for the section of Eurocode 1 on fire actions.

Venkatesh Kodur is a Professor in the department of Civil & Environmental Engineering and also serves as Director of the Center on Structural Fire Safety and Diagnostics at the Michigan State University (MSU). Before moving to MSU, he worked as a senior Research Officer at the National Research Council of Canada (NRCC), where he carried out research in structural fire safety field. He received his M.Sc. and Ph.D. from Queen's University, Canada in 1988 and 1992, respectively. He currently teaches under graduate and graduate courses which include structural fire engineering. He directs the research of 1 PDF, 7 PhD, 7 MS, and several undergraduate students.

Dr. Kodur's research has focused on the evaluation of fire resistance of structural systems through large scale fire experiments and numerical modelling; characterization of the materials under high temperature (constitutive modeling); performance based fire safety design of structures; and non-linear design and analysis of structural systems. He has collaborated closely with various industries, funding agencies, and international organisations and developed simplified design approaches for evaluating fire resistance, and innovative and cost-effective solutions for enhancing fire-resistance of structural systems. Many of these design approaches and fire resistance solutions have been incorporated in to various codes and standards. He has published over 175 peer-reviewed papers in international journals and conferences in structural and fire resistance areas and has given numerous invited key-note presentations.

Dr. Kodur is a professional engineer, Fellow of ASCE and Fellow of ACI and member of SFPE and CSCE. He is also an Associate Editor of Journal of Structural Engineering, Chairman of ASCE-SFPE Standards Committee on Structural Design for Fire Conditions, Chairman of ACI-TMS Committee 216 on Fire Protection and a member of EPSRC (UK) College of Reviewers. He has won many awards including AISC Faculty Fellowship Award for innovation in structural steel design and construction (2007), NRCC outstanding achievement award (2003) and NATO award for collaborative research. Dr. Kodur was part of the FEMA/ASCE Building Performance Assessment Team that studied the collapse of WTC buildings as a result of September 11 incidents.

Raul Zaharia is Associate Professor at the Politehnica University of Timisoara, Romania. He graduated from the same university in 1993 and made his Ph. D. thesis in the field of steel structures. In 2000, he was awarded a one year postdoctoral grant at the University of Liege by the Services of the Prime Minister of Belgium for Scientific, Technical and Cultural Affairs, for research works in the field of fire design. During this period, he studied the behaviour of high-rise steel rack structures in fire and started the collaboration with Jean Marc Franssen in the field of fire design. Other experiences abroad include periods spent at CTICM France, ELSA Italy and City University of London for a total of three years. His experience in the field of steel structures and fire design involves research, reflected by more than 80 scientific papers and research reports, but also advanced fire calculation for some buildings built in Bucharest, Romania. At the University of Timisoara, he initiated the master course "Fire design of civil engineering structures". He participated to the translation of the fire parts of the EN1993 for steel structures, EN1994 for steel concrete structures and EN1999 for aluminium structures in Romanian for ASRO (Romanian Association for Standardisation) and was part of the team which elaborated the Romanian National Annexes of these documents. He is member of the ECCS TC3 Committee "Fire Design" as expert from Romania.

Chapter 1

Introduction

1.1 Fire safety design

Fire represents one of the most severe environmental hazards to which buildings and built-infrastructure are subjected, and thus fires account for significant personal, capital and production loss in most countries of the world each year. Therefore, the provision of appropriate measures for protecting life and property are the prime objectives of fire safety design in buildings. These fire safety objectives can be achieved through different strategies. During the design process, and in the initial stages of a fire, the first aim is to confine the fire inside the compartment so that it does not spread to other parts of the building. However, if the fire becomes large despite of preventive measures, then the aim of design is to ensure that the building remains structurally stable for a period of time enough to evacuate the occupants and for the firefighters to contain the fire. It is the job of the designer to ensure the effectiveness of the chosen measures in preventing the fires from spreading and destroying the building.

Measures to retard fire growth include use of low fire hazard materials, using fire protection, the use of Sprinklers, and provisions to facilitate fire department operations. In addition to these measures, there should be careful control over the combustible materials that are brought into a building on a regular basis as part of the function of the structure, i.e. warehouse for fuels, residence, storage areas, etc.

To protect people, measures should be taken against the hazards of the spread of fire and its combustion products. The design of the building should ensure safety of the people in case of fire by providing adequate exit avenues and time, preventing the spread of smoke and hot gases, and ensuring the integrity of the structure for a reasonable time period under fire.

The closest measures related to building design are those related to fire confinement. These measures include ensuring sufficient time of structural integrity and stability for evacuation of people and extinguishing the fire, and providing fire barriers capable of delaying or preventing the spread of fire from one room to another.

1.2 Codes and standards

1.2.1 General

Methods, experiences, and related measures for achieving the maximum safety against fires, are usually found in related national codes and standards. The given methods and measures in these codes and standards, affect the design strategies of the structure for maximizing the safety against fires.

Codes are usually more general than standards, and they include in their framework references to standards. Model codes consist of comprehensive building regulations suitable for adoption as law by municipalities and it establishes basic pattern for building codes throughout the country. Codes, when adopted by state and municipalities, are intended to become regulated through legislation. Generally, building codes specify minimum requirements for design and construction of buildings and structures. These minimum requirements are established to protect health and safety of the public and generally represent a compromise between optimum safety and economic feasibility. Structural design, fire protection, means of egress, light, sanitation, and interior finish are general features covered in the building codes.

Standards are one step below the codes and are considered to be a set of conditions or requirements to be met by a material, product, process, or procedure. Also, standards give detailed method of testing to determine physical, functional, or performance characteristics of materials or products. Usually, standards are referenced in building codes, thus keeping the building codes to a workable size. Also, standards are used by specification writers in the design stage of a building to provide guidelines for bidders and contractors.

Standards referenced in building codes can be generally classified into materials standards, engineering practice standards and testing standards.

Materials standards generally establish minimum requirements of quality as measured by composition, mechanical properties, dimensions and uniformity of product. They include methods of sampling, handling, storage and testing for verification of such quality. Engineering practice standards include basic design procedure engineering formulas and specified provisions intended to provide satisfied engineering performance. Testing standards describe procedures for testing quality and performance of materials or assemblies. They include procedures for testing and measuring characteristics such as strength, stability, durability, combustibility and fire resistance.

1.2.2 Fire safety codes

The fire safety requirements for buildings are generally contained in national building codes or fire codes. Current practice for fire safety design of structures in most countries is principally based on the provisions of locally adopted codes that are usually based upon the model building codes. The codes specify minimum required fire endurance times (or fire endurance ratings) for building elements and accepted methods for determining their fire endurance ratings.

Buildings codes, based on the prescribed provisions, can be classified into: prescriptive codes and performance codes. In the prescriptive codes, full details about what materials can be used, the maximum or minimum size of building, and how components should be assembled are specified. In the performance codes the objectives to be met, and the criteria to be followed to achieve the set-out objectives are listed. Therefore, freedom is allowed for the designer and builder for selecting materials and methods of construction as long as it can be shown that the performance criteria can be met. Performance codes still include fair amount of specification-type requirements, but special attention is given for provisions allowing alternate methods and materials if proven adequate.

The prescriptive approach to fire safety engineering is simple to implement and enforce, and given the lack of documented structural failure while in use, has been deemed satisfactory in meeting the codes' stated intent, which is "...*to establish the minimum requirements to safeguard the public health, safety and general welfare through structural strength, means of egress facilities, stability, sanitation, adequate light and ventilation, energy conservation, and safety to life and property from fire and other hazards attributed to the built environment.*" However, there is also a growing recognition that the current prescriptive, component-based method only provides a relative comparison of how similar building elements performed under a standard fire exposure, and does not provide information about the actual performance (i.e. load carrying capacity) of a component or assembly in a real fire environment (SFPE, 2000). Thus, for a certain class of buildings such as high-rises or other important structures, which, due to the longer evacuation time or the significance of the buildings, may be required to survive beyond the maximum code required fire endurance time without structural collapse, a performance-based fire resistance approach may provide a more rational method for achieving the necessary fire resistance that is more consistent with the needed level of protection. A performance based fire resistance approach considering the evolution of the building's structural capacity as it undergoes realistic (non-standard) fire exposures is thus a desirable alternative fire resistance design method for those structures.

1.2.3 North American codes and standards

In the United States there are two main building codes; the International Building Code (IBC, 2006) and the NFPA Building and Construction Code (NFPA, 2003). In Canada, National building Code of Canada (NBCC) contains the requirements for building design. Both ICC and NFPA codes specify fire resistance ratings in a prescriptive environment. However, the latest edition of NBCC (2005) contains objective based provisions for fire safety design. Designers have the freedom to select materials and assemblies to meet these pre-required requirements using the methods specified in referenced standards. Moves towards performance based codes are being taken slowly in US codes (ICC 2003, ICC 2003c, NFPA 2003). NFPA (2003) contains some specific provisions for undertaking performance based fire safety design. ICC, NFPA and NBCC codes reference nationally accepted standards which contain permitted methods for determining fire endurance ratings for various material types. This includes standards for fire testing, standards for undertaking structural (and fire) design of steel structures and general standards or manuals for fire resistance design.

The loads to be considered in design of buildings are specified in ASCE-07 (2005). This standard contains different loads and load combinations that are to be included in computing design forces on a structures. The load requirements in Canada are part of National Building Code of Canada (2005). In US the specifications for testing structural elements under fire exposure is to be in accordance with procedure and criteria set forth in ASTM E 119 (2005) or NFPA 251(2006) or UL 263 (2003). All these standards have similar specifications and are considered equivalent in most situations. In Canada, the fire test provision for structural members is specified in ULC/CSA S101 standard

(2004). Many of the provisions in these standards are similar to those of ISO 834 (2002).

In US, AISC (American Institute of Steel Construction) steel construction manual (AISC, 2005) is the principal reference that contains documentation for design of steel structures. Recent edition of the AISC Manual contains both LRFD and ASD specifications for the design of steel structures. A general discussion on overall fire design is provided in this manual. However, there is very limited information on the fire design provisions for steel structures. AISC Manual gives some information on thermal and mechanical properties at elevated temperatures, which will allow the design of single members for fire exposure. A recent report by AISC (Rudy et al. 2003) gives a survey of existing codes and standards, plus a lot of background information on fire testing, analysis and design methods for steel structures.

In addition to the AISC manual and guides, ASCE/SFPE 29 contain a number of analytical methods for determining equivalent fire resistance ratings for steel structural members. Most of these analytical methods have been developed based on the results of standard fire resistance tests carried out on steel structural assemblies under standard fire exposure. Another source for fire resistance calculation methods in US is the SFPE Handbook of Fire Protection Engineering (SFPE, 2002) which has a chapter on steel design that gives an overview of performance of steel structures in fire, but this does not give sufficient information for the advanced calculation methods as in Eurocode 3(2003). SFPE (2002) also gives a chapter on high temperature material properties of steel and insulation materials at elevated temperature. National building code of Canada (NBCC 2005) contains a number of analytical methods and a number of tabulated data for determining fire resistance ratings for steel structural members.

1.2.4 European codes: the Eurocodes

Provisions relating to design and analysis of various structural systems used in buildings are detailed in Eurocodes. The structural Eurocodes form a set of documents for determining the actions and for calculating the stability of building constructions.

- EN 1990 defines the general rules governing the ultimate limit state design which is the basic philosophy of the Eurocodes.
- EN 1991 gives the design values of the actions.
- Eurocodes 2 to 6 and Eurocode 9 deal with the design of structures made of different materials: EN 1992 is for concrete structures, EN 1993 is for steel structures, EN 1994 is for composite steel-concrete structures, EN 1995 is for timber structures, EN 1996 is for masonry structures and EN 1999 is for aluminium structures.
- Finally, EN 1997 is especially dedicated to geotechnical design while EN 1998 covers earthquake resistance.

Each Eurocode is designated by a number in the CEN classification, starting from 1990 for the basis of design to 1999 for aluminium alloy structures. It has to be noted that these numbers have nothing to do with a date of publication or whatsoever. It is just fortuitous that these numbers look like numbers of years that coincide with the period when these documents have been published.

In the period before the Eurocodes were adopted by CEN and became EN documents, they were simply referred to as, for example, *Eurocode 1* or *Eurocode 5*. It is quite fortunate that the last digit of the numbers in the CEN designation is the same as the number of the corresponding Eurocode. For example, EN 1993 is Eurocode 3.

1.3 Design for fire resistance

Current fire protection strategies for building often incorporate a combination of active and passive fire protection systems. Active measures, such as fire alarms and detection systems or sprinklers, require either human intervention or automatic activation and help control fire spread and its effect as needed at the time of the fire. Passive fire protection measures are built into the structural system by choice of building materials, dimensions of building components, compartmentation, and fire protection materials, and control fire spread and its effect by providing sufficient *fire resistance* to prevent loss of structural stability within a prescribed time period, which is based on the building's occupancy and fire safety objectives. Materials and construction assemblies that provide fire resistance, measured in terms of fire endurance time, are commonly referred to as *fire resistance-rated-construction* or *fire-resistive materials and construction* in the model building codes. Recent advances in fire science and mathematical modeling has led to the development of rational approaches for evaluating fire resistance. Therefore it is possible to design for required fire resistance of structural members.

1.3.1 Fire resistance requirements

Fire resistance of a building components or assembly can be defined as its ability to withstand exposure to fire without loss of load bearing function, or act as a barrier against spread of fire, or both. Fire resistance is expressed as the length of time period that a construction can withstand exposure to standard fire without losing its load bearing strength or fire separating function. This time is a measure of the fire performance of the structure and is termed the "fire resistance" of that structure, and it is widely used in most building codes and material standards. In North American standards the term 'fire endurance' is often used to describe both load bearing and fire separation function for structural assemblies.

The required fire resistance ratings for building components are specified in building codes. Fire resistance ratings denote the required fire resistance rounded off to the nearest hour or half hour. These fire resistance requirements are function of several factors such as fire loads, building occupancy, height, and area. In reality, fire resistance is a function of many other factors that are not considered in the building codes, such as the properties of the material of the wall enclosing the fire, dimensions of the openings, and heat lost to the surroundings.

A major difference between the standard fire temperature curve and an actual fire temperature curve is that the standard fire curve keeps on increasing with time, whereas the actual fire usually decreases after reaching a certain maximum. The standard fire curves in most countries are very close to that of the ISO 834 standard. The use of actual fire temperature curves would give more accurate information on the fire performance of the construction. However, the current practice still uses the standard fire temperature curve to express fire resistance. All the provision and ratings in North

American codes and standards are based on exposure to the standard fire. Therefore, there is a large amount of information on the standard fire resistance of numerous building components and assemblies.

1.3.2 Fire resistance assessment

A very common method to assess fire resistance is through testing specimens, such as beams, columns and walls, or assemblies under fire. Generally, fire resistance has been established through laboratory tests in accordance with procedures specified in standards such as ISO 834 or ASTM E119. These test methods are used to evaluate standard fire resistance of walls, beams, columns, floor, and roof assemblies. In addition to ISO and ASTM procedures, other organizations also publish fire test procedures which are virtually identical to those developed by the ISO and ASTM, such as the NFPA, and Underwriters Laboratories, Inc. "UL", Underwriters Laboratories of Canada "ULC", and the standards Council of Canada. In all of these procedures, fire resistance is expressed in term of exposure to a standard fire. Generally, there are three criteria in a standard test method. Namely, load-bearing capacity, integrity, and the temperature rise on the unexposed side for fire barriers.

Most countries around the world rely on large scale fire resistance tests to assess the fire performance of building materials and structural elements. The time temperature curve used in fire resistance tests is called the *standard fire*. Full size tests are preferred over small scale tests because they allow the method of construction to be assessed, including the effects of thermal expansion, shrinkage, local damage and deformation under load.

Advancement in theoretical prediction of fire resistance has been rapid in recent years. In many cases, the fire resistance of building components can also be determined through calculations. Calculation methods are less expensive and time consuming than conducting fire resistance tests, which are usually carried out for large scale test specimens.

In recent years the calculation methods are slowly being incorporated into various codes and standards. North American Codes are slow in incorporating calculation methodologies and only contain simple equations based on prescriptive approaches. There is very little guidance on sophisticated analysis or design from first principles. However, Eurocodes are much more progressive in adopting calculation methods for evaluating fire resistance. The fire design provisions in Eurocodes are well received by the engineering community and are often referenced in National Codes and Standards worldwide. An example of this is the recent edition of AISC Steel design manual (2005) which contains reference to Eurocode 3 for fire resistance provisions.

1.3.3 Eurocodes

Except for the Eurocodes dealing with the bases of design, geotechnical design and earthquake resistance, each Eurocode is made of several parts, including part 1-1 that covers the general rules for the design at ambient temperature and part 1-2 that covers the design in the fire situation.

Table 1.1 presents a list of the different Eurocodes and Figure 1.1, from Gulvanessian et al. 2002, gives a synoptic view of these documents and how these different codes relate to each other.

Table 1.1 Classification of the Eurocodes

Eurocode number	Ambient conditions	Fire conditions
Basis of design	EN 1990	–
Actions	EN 1991-1-1	EN 1991-1-2
Concrete structures	EN 1992-1-1	EN 1992-1-2
Steel structures	EN 1993-1-1	EN 1993-1-2
Composite steel-concrete structures	EN 1994-1-1	EN 1994-1-2
Timber structures	EN 1995-1-1	EN 1995-1-2
Masonry structures	EN 1996-1-1	EN 1996-1-2
Geotechnical design	EN 1997	–
Earthquake resistance	EN 1998	–
Aluminium alloy structures	EN 1999-1-1	EN 1999-1-2

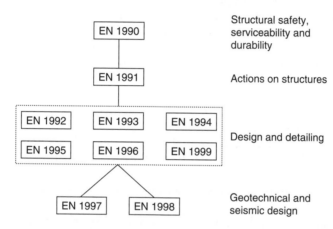

Fig. 1.1 Synoptic view of the different Eurocodes

Because this book is on structures subjected to fire, any reference to a Eurocode without any further qualification will imply that the fire part of the corresponding Eurocode is referred to. If required, the distinction will be made between *the cold Eurocode*, i.e. part 1-1, and *the fire Eurocode*, i.e. part 1-2.

Historically speaking, two different stages of development have to be recognised in the CEN designation: the stage when the Eurocodes had the status of provisory documents, called the ENV stage, and the stage when they became final documents, called the EN stage. When no reference is made to the particular version of a Eurocode, either ENV or EN, it will simply be called "*the Eurocode*" or "*Eurocode 3*" in this book, sometimes noted as "*EC3*". When it is necessary to mark the distinction between the provisory and the final stage, the terms "*ENV*" or "*EN*" will be used, with reference to the preliminary drafts if required, "*prEN*" or "*prENV*".

1.3.4 *Scope of Eurocode 3 – Fire part*

Eurocode 3 does not deal with the insulation or the integrity criteria of separating elements (criteria E and I). If a separating wall is made of steel sandwich panels, for example, it is nowadays not possible to calculate or to simulate its behaviour under fire. This is because the behaviour of such a wall involves several complex phenomena that cannot be predicted or modelled at the time being such as, for example, the opening of joints between adjacent panels, the shrinkage of insulating materials, the movements of the insulating material that creates some gaps and air layers between the material and the steel panel, the high thermal deformations of the steel sheets, the local influence of fastening elements, etc. Such separating elements have to be tested experimentally. Should new developments be undertaken in the direction of modelling separating walls based on steel elements, the thermal and mechanical properties of steel given in the Eurocode could serve as a starting point.

Eurocode 3 deals essentially with the load bearing capacity of steel elements and structures, i.e. the mechanical resistance (criterion R). It gives information that allows calculating whether or how long a steel structure is able to withstand the applied loads during a fire. The design is thus performed in the ultimate limit state (see Chapter 2).

There is strictly speaking no deformation criteria explicitly mentioned in the Eurocode such as, for example, a limit equal to 1/30 of the span of a simply supported beam as is found in different standards for experimental testing. Deformation control is anyway mentioned in the basis of design. Deformation criteria should be applied in two cases:

1. When the protection means, for example the thermal insulation protecting the steel member, may loose its efficiency in case of excessive deformations.
2. When separating elements, for example a separating wall, supported by or located under the steel member may suffer from excessive deformations of this member.

The above criterion can be exempted under the following two cases.

1. When the efficiency of the means of protection, either a thermal insulation applied on the section or a false ceiling under a beam, has been evaluated using the test procedures specified in EN 13381-1, EN 13381-2 or 13381-4 as appropriate. The rational for this exception is that these test procedures comprise at least one test on a loaded member and the effects of eventual deformations of the member are thus implicitly taken into account into the equivalent thermal properties of the insulating material or in the shielding effect provided by the false ceiling.
2. When the separating element has to fulfil requirements according to a nominal fire exposure. The rational for this exception is that a nominal fire is an arbitrary fire exposure that allows comparing different construction systems between each other. It is by no means a representation of what could occur to the system in a real fire. It would then not be consistent trying to estimate deformations of the structural elements and to compare it to deformation criteria if the starting point of the design is purely arbitrary. According to an expression recommended by Buchanan, there is a need of consistent level of crudeness.

Advanced calculation models (see Chapter 7) automatically provide the deformations of the structure. These deformations can be compared with any defined deformation criteria. It has to be noted, however, that the Eurocode does not provide the deformation criteria that have to be used. The designer has to concur with the authority having jurisdiction on the limiting criteria to be applied for evaluating failure. It should also be checked that failure will not occur from excessive deformations causing one of the member of the structure falling from its supports. Finally, the deformations at the ultimate limit state implied by the calculation method should be limited to ensure that compatibility is maintained between all parts of the structure.

Simple calculation models, on the other hand, do not yield the deformation of the structure at the ultimate limit state. It is thus not possible to comply with the requirement given in the Eurocode on deformation criteria when using simple calculation models. In fact, for a Class 1 or Class 2 beam in bending, for example (see Section 5.5) plastic theory shows that the plastic moment in a section can be obtained only for an infinitely small radius of curvature, i.e. for an infinite deformation.

In practice, deformation criteria are normally ignored when using simple calculation models. If a particular situation leads the designer to believe that a real attention should be paid to the deformation, a practical way to limit the deformations is to use the 0.2 percent proof strength instead of the effective yield strength (see Section 5.6.6.)

1.4 General layout of this book

The information that is necessary to perform the fire design of a structure made of a particular material, say a steel structure, is:

(a)   The basis of design, stated in EN 1990.
(b)   The mechanical actions, i.e. the forces, acting on the structure in the fire situation. Some information is found in EN 1991-1-2, but explicit reference is also made to EN 1991-1-1 that is therefore also necessary.
(c)   The thermal actions, i.e. the fire and the heat flux induced in the elements by the fire. The information is in EN 1991-1-2.
(d)   The rules for determining the temperatures in the structure during the course of the fire. They are given in the fire part of the relevant material Eurocode, e.g. EN 1993-1-2 for a steel structure.
(e)   The rules for determining the structural stability. They are given in the fire part of the relevant material Eurocode, e.g. EN 1993-1-2 for a steel structure, but reference is often made to the cold part of the same material Eurocode, EN 1993-1-1 for a steel structure.

This layout is valid in general but some exceptions do exist: the structural stability of timber elements, for example, does not necessarily require the determination of the temperatures in the element and point d) is not required. The same holds if the fire resistance is determined from tabulated data, for concrete elements for example.

The rest of this book is organised according to the following layout, illustrated by Figure 1–2:

Chapter 2 deals with basis of design and mechanical loads.
Chapter 3 deals with thermal response from the fire.
Chapter 4 deals with thermal analysis by simple calculation models.
Chapter 5 deals with mechanical analysis by simple calculation models.
Chapter 6 deals with the design of joints.

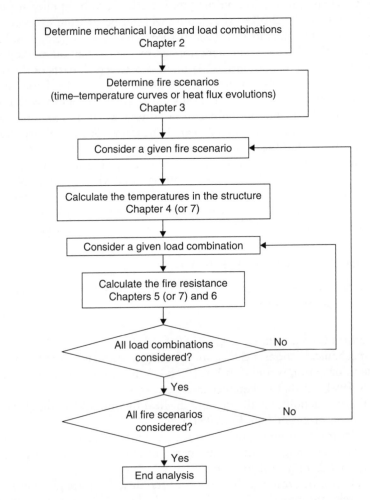

Fig. 1.2 General layout of an analysis

Chapter 7 deals with thermal and mechanical analysis by advanced calculation models. Chapter 8 gives four design examples showing how a complex structure can be designed using the concept of element or substructure analysis.

Chapter 2

Mechanical Loading

2.1 Fundamental principles

2.1.1 Eurocodes load provisions

The design philosophy of the Eurocodes is based on the concept of limit states, i.e. states beyond which the structure no longer satisfies the design performance requirements. The Eurocodes treat the fire exposure to be an accidental situation and this requires only verification against the ultimate limit state (as opposed to the serviceability limit state). Ultimate limit states is associated with structural collapse or other similar forms of structural failure such as loss of equilibrium, failure by excessive deformation, formation of a mechanism, rupture or loss of stability.

In the semi probabilistic approach, the design against the ultimate limit state is based on the comparison between the resistance of the structure calculated with the design values of material properties, on one hand, and the effects of actions calculated with design value of actions, on the other hand. This is represented as:

$$R_{fi,d,t}(X_{d,fi}) > E_{fi,d}(F_{fi,d})$$ (2.1)

where:

$R_{fi,d,t}$ is the design value of the resistance in case of fire,
$X_{d,fi}$ is the design value of the material properties in case of fire,
$E_{fi,d}$ is the design value of the effects of actions in case of fire,
$F_{fi,d}$ is the design value of the actions in case of fire.

The resistance and the effects of actions are both based on characteristic values of geometrical data, usually the dimensions specified in the design, for cross section sizes for example. Geometrical imperfections such as bar out of straightness or frame initial inclinations are represented by design values.

The design values of the material properties, $X_{d,fi}$, are described for each material in the relevant material Eurocode, for example in Eurocode 3 for steel structures. These material Eurocodes also describe how the resistance, $R_{fi,d,t}$, based on these material properties, is calculated. Eurocode 1 describes how the design values of actions, $F_{fi,d}$, are calculated.

The partial factor method considers that design values are derived from representative, or characteristic, values multiplied by scalar factors. The general equations are:

$$G_{fi,d} = \gamma_G G_k \qquad \text{for the permanent actions} \qquad (2.2)$$

$$Q_{fi,d} = \gamma_Q Q_k, \gamma_Q \psi_0 Q_k, \psi_1 Q_k \text{ or } \psi_2 Q_k, \quad \text{for the variable actions} \qquad (2.3)$$

$$P_{fi,d} = \gamma_P P_k \qquad \text{for the prestressing actions} \qquad (2.4)$$

where:

G_k, Q_k, P_k	are the characteristic values of the permanent, variable and prestressing action,
$G_{fi,d}, Q_{fi,d}, P_{fi,d}$	are the design values of these actions in case of fire,
$\gamma_G, \gamma_Q, \gamma_P$	are the partial factors for these actions,
ψ_0	is the coefficient for combination value of a variable action, taking into account the reduced probability of simultaneous occurrences of the most unfavourable values of several independent actions,
ψ_1	is the coefficient for frequent value of a variable action, generally representing the value that is exceeded with a frequency of 0.05, or 300 times per year,
ψ_2	is the coefficient for quasi-permanent value of a variable action, generally representing the value that is exceeded with a frequency of 0.50, or the average value over a period of time.

Different actions generally occur simultaneously on the structure. In an accidental situation, they have to be combined as follows:

- Design values of permanent actions
- Design value of the accidental action
- Frequent value of the dominant variable action
- Quasi-permanent values of other variable actions.

When it is not obvious to determine which one amongst the variable actions is the dominant one, each variable action should be considered in turn as the dominant action, which leads to as many different combinations to be considered.

In case of fire, which is an accidental design situation, and if the variability of the permanent action is small (applicable in most cases), the following symbolic equations Equation 2.5a or Equation 2.5b hold:

$$E_{fi,d} = G_k + P_k + \psi_{1,1} Q_{k1} + \sum_{i>1} \psi_{2,i} Q_{ki} \qquad (2.5a)$$

$$E_{fi,d} = G_k + P_k + \sum_{i=1} \psi_{2,i} Q_{ki} \qquad (2.5b)$$

It may be noticed that partial factors for permanent, prestressing and variable actions are equal to 1.0 in accidental situation.

Table 2.1 Coefficients for load combination factors ψ for buildings

Action	ψ_1	ψ_2
Imposed load in buildings		
category A: domestic, residential	0.5	0.3
category B: offices	0.5	0.3
category C: congregation areas	0.7	0.6
category D: shopping	0.7	0.6
category E: storage	0.9	0.8
Traffic loads in buildings		
category F: vehicle weight $\leq$ 30 kN	0.7	0.6
category G: 30 kN < vehicle weight < 160 kN	0.5	0.3
category H: roofs	0.0	0.0
Snow loads		
for sites located at altitude H $\leq$ 1000 m	0.2	0.0
for sites located at altitude H > 1000 m	0.5	0.2
Wind loads	0.2	0.0

Table 2.1 given here is table A1-1 of prEN 1990 and gives the relevant ψ factors for the fire situation in buildings.

The choice whether the frequent value (Eq. 2.5a) or the quasi-permanent value (Eq. 2.5b) has to be used for the dominant variable action is a nationally determined parameter. In this book, Equation 2.5a will normally be used because it leads to the most complete load combinations and is thus more illustrative for the examples.

In fact, Equation 2.5a was the only one mentioned in ENV 1991-1-2. Equation 2.5b appeared in prEN 1991-1-2 and, in EN 1991-1-2, the use of the quasi-permanent value is recommended for variable actions. The motivation to change from the frequent to the quasi-permanent value when the ENV was changed into prEN was that this is the solution used for earthquake, which is also an accidental action, just as the fire.

Why should the fire be treated differently? The authors of this book think that this argument can be accepted, except for the wind action. Indeed, the coefficient for the quasi-permanent action ψ_2 for wind is 0 and if wind is taken with a 0 value even when it is the dominant variable action, there will be no verification at all under horizontal forces for a structure submitted to the fire. In case of an earthquake, horizontal forces from the accidental action are of course always present and the wind effect may not be of significant importance anyway.

In fact, not only the choice between Ψ_1 or Ψ_2 is a nationally determined parameter but also the values of these factors. Values different from those presented in Table 2.1 can be adopted in different countries.

The design value of the accidental action that was mentioned above does not appear in Equation 2.5 because, in case of fire, the fire action is not of the same form as the other actions. It does not consist of some N or some N/m^2 that could be added to the dead weight or to the wind load. The fire action consists of indirect effects of actions induced in the structure by differential and/or restrained thermal expansion. Whether and how these effects have to be taken into account is discussed in the different material Eurocodes.

2.1.2 American provisions for fire design

US model codes use ASCE-07 (ASCE, 2005) as the reference standard for various loads to be considered in the design of buildings. ASCE/SEI 07 contains detailed specifications for evaluating the loads under various actions. This standard utilizes the well accepted principle that the likely loads that occur at the time of a fire are much lower than the maximum design loads specified for room temperature conditions. This is especially true for members which have been designed for load combinations including wind, snow or earthquake. For this reason, different design loads and load combinations are used. It is generally assumed that there is no explosion or other structural damage associated with the fire. Loads on members could be much higher if some members are removed or distressed.

In ASCE-07 fire is considered as an extraordinary event. A statement in Section C2.5 of the standard states "*For checking the capacity of a structure or structural element to withstand the effect of an extraordinary event, the following load combination should be used*": Accordingly, the design load combination for fire (U_f) is given as:

$$U_f = 1.2D_n + 0.5L_n \tag{2.6}$$

where D_n and L_n are the design levels of dead and live load respectively, from the standard.

The computed loads as per this provision generally works out be lower than the maximum design loads on the structure, especially for members sized for deflection control or architectural reasons.

In Canada, the load provisions are specified in National Building Code of Canada (NBCC 2005) and the provisions for loads under fire conditions is similar to those in ASCE -07.

2.2 Examples

2.2.1 Office building

What are the relevant load combinations for an office building that is not submitted to traffic loads and has no prestressed concrete element (H < 1000 m)?

If Equation 2.5a is used, the appropriate values of ψ from Table 2.1 yield the following combinations for the applied loads:

- Live load is the dominant variable action.

$$E_{fi,d} = \text{dead weight} + 0.5 \times \text{live load} \tag{2.6}$$

- Loading due to snow is the dominant variable action.

$$E_{fi,d} = \text{dead weight} + 0.2 \times \text{snow load} + 0.3 \times \text{live load} \tag{2.7}$$

- Load from wind is the dominant variable action.

$$E_{fi,d} = \text{dead weight} + 0.2 \times \text{wind load} + 0.3 \times \text{live load} \tag{2.8}$$

If Equation 2.5b is used, only one combination has to be considered:

$$E_{fi,d} = \text{dead weight} + 0.3 \times \text{live load} \tag{2.9}$$

The wind load may take different patterns and values, depending on the direction of the wind and whether the wind induces a positive or a negative pressure inside the building. This can increase significantly the amount of calculations but, especially in sway frames or for the bracing system of a laterally supported structure, this cannot be avoided.

2.2.2 Beam for a shopping centre

What is the design load on a beam that is part of a floor in a shopping centre?

A beam supporting a floor in this type of building is designed using Equation 2.10 (from Eq. 2.5a) because neither wind nor snow can affect such a beam.

$$E_{fi,d} = \text{dead weight} + 0.7 \times \text{live load} \tag{2.10}$$

2.2.3 Beam in a roof

What is the design load supported by a beam that is part of a roof (H > 1000 m)?

A beam supporting the roof of a building is designed using Equation 2.11 if snow is the dominant variable action and Equation 2.12 if wind is the dominant variable action.

$$E_{fi,d} = \text{dead weight} + 0.5 \times \text{snow load} \tag{2.11}$$

$$E_{fi,d} = \text{dead weight} + 0.2 \times \text{wind load} + 0.2 \times \text{snow load} \tag{2.12}$$

2.3 Specific considerations

2.3.1 Simultaneous occurrence

Clause 4.2.2 (1) of EN 1991-1-2 says that "*Simultaneous occurrence with other independent accidental actions needs not be considered*". The important word is *independent*. A fire and a tornado can be considered as independent and they will not be combined. An earthquake, on the other hand, frequently gives rise to numerous building fires. In this case, the actions are not really independent. Whether the combination or, better, the succession of these two events has to be considered should be decided on a case per case basis in consultation with the authorities or the contractor. This could also be the case for a fire resulting from a terrorist action such as a bombing, or impact resulting from a vehicle hitting the building in an accident. Designing a structure that would be able to sustain successively two of these accidental actions is of course not without an influence on the cost of the structure. In this book, it is considered that the fire is the only accidental action.

2.3.2 Dead weight

The dead weight must be considered in all loading scenarios. It is important that all components of the dead weight are considered. For example, in a residential building, this would not be a significant approximation to neglect the lightning devices in the verification of the concrete floors. On the other hand, in a lightweight steel industrial building, the dead weight of venting ducts suspended to the beams of the roof or the supporting beams of a crane can form an important part of the dead weight. This is also the case for equipments or reservoirs located on and supported by the roof.

2.3.3   Upper floor in an open car park

In an open car park with access for vehicles on the upper floor, it does not seem realistic to combine the traffic load and the snow load. This floor is calculated under the two different hypotheses:

1.   With the traffic load. It is then not considered as the roof of the building but as a floor.
2.   With the snow load and no traffic.

The snow load on the upper floor and the traffic loads on the other floors can of course be combined.

2.3.4   Industrial cranes

For moving lifting cranes in industrial buildings, the dead weight of the longitudinal supporting beams spanning from one frame to the others has to be considered, as stated previously.

This is also the case for the transverse beams on which the motor is rolling. The weight of the motor is supported equally by these two transverse beams. The motor may be located near the column of the frame with the highest compression force. The longitudinal position of the crane is chosen in such a way that it maximise the effects on one of the frames, see Figure 2.1.

It is current practice not to consider any load suspended to the crane because, even if there was one at the beginning of the fire, the cable is a rather thin element that is likely to heat up faster than any part of the building structure with the consequence that the load might rapidly lay down on the ground. Operational instruction can also be given to the personnel operating the crane to lay down any load on the floor before evacuation of the building in case of fire. Whether this can be safely relied on can be a subject of discussion. In special cases, it might be desirable to gather additional information on the statistical distribution in time of the value of this load and to deduce a design value in case of fire.

It has to be noted that EN 1991-1-2 states explicitly under clause 4.2.1 (5) that "*Loads resulting from industrial operations are generally not taken into account*". This can also be seen as a justification for not taking any load supported by the crane into account

2.3.5   Indirect fire actions

Indirect fire actions are defined in Eurocode 1 as internal forces and moments caused by thermal expansion. Further in the text, in Clause 4.1 (1) of EC1, it is recognised that they result from imposed and constrained expansions and deformations.

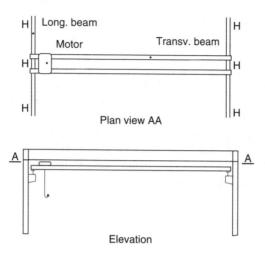

Fig. 2.1 Typical crane in an industrial building

They shall be considered apart from those cases where they:

(a)   may be recognized a priori to be either negligible or favourable;
(b)   are accounted for by conservative support and boundary conditions and/or conservatively specified fire safety requirements.

Some engineering judgement is thus necessary to decide in each particular case whether at least one of these conditions is met.

A particular case is quoted under Clause 4.1 (4): "…*when fire safety requirements refer to members under standard fire conditions*". This is the case, for example, if the requirement is expressed as 60 minutes fire resistance to the standard fire for the columns and 30 minutes for the beams. The motivation behind this permission is probably to be found in the fact that, historically, requirements made on members and based on the standard fire were linked to a verification by an experimental fire test in which no indirect actions were present. If a calculation model is used to verify such an arbitrary requirement, this is with the objective to obtain a result similar to the result that a fire test would have given, but at a lower cost and much faster. The objective is not to obtain a representation of the real fire behaviour of the structure. The calculation model must therefore represent as close as possible the conditions of the test and, hence, no indirect action is taken into account.

If the requirement refers to the entire structure as a whole, or if the requirement refers to anything else that the standard fire, a fire model for example, it does not necessarily mean that the indirect actions have automatically to be taken into account, but authorisation not to do so is not automatic.

If indirect fire actions need not be considered, either because of the permission stated in Clause 4.1 (4) or because it is considered that one of the conditions a) and b) given previously is met, then the effects of actions are constant throughout the fire exposure and may thus be determined at time $t = 0$.

The question of indirect fire actions will be discussed further in Chapter 5, when the analysis of substructures is introduced.

2.3.6 Simplified rule

EN 1991-1-2 states in Clause 4.2.1 (1)P that "*Actions shall be considered as for normal temperature design, if they are likely to act in the fire situation*". In other words, actions acting in the fire situation should be considered in the design of the structure subjected to the fire. It goes without saying, one might think. Yet, this sound principle is contradicted by a simplified rule. According to clause 4.3.2 (2), in the cases where indirect fire actions need not be explicitly considered, "*effects of actions may be deduced from those determined in normal temperature design*" by a multiplication factor η_{fi}, according to Equation 2.13.

$$E_{fi,d,t} = \eta_{fi} E_d \tag{2.13}$$

where:

E_d is the design value of the relevant effects of actions from the fundamental combination according to EN 1991-1-1,

η_{fi} is a reduction factor,

with $\eta_{fi} = (G_k + \psi_{fi} Q_{k,1})/(\gamma_G G_k + \gamma_{Q,1} Q_{k,1})$

and $\psi_{fi} = \psi_{1,1}$ or $\psi_{2,1}$ depending on the choice made for the nationally determined parameter, see § 2.1.

The reduction factor η_{fi} is smaller than 1.0 and stems from the reduction of the design loads when going from the room temperature situation to the fire situation. The idea seems to be interesting if the effects of actions at room temperature have been determined in a complex structure by a manual method, the "method of Cross" for example. It is in this case considered as a benefit to be allowed to simply multiply these results by a scalar factor rather than doing one or several additional analyses for the fire situation.

The authors of this book nevertheless do not recommend the utilisation of this "simplified" rule but rather advocate determining the effects of action in case of fire with the basic Equation 2.5. This is because:

1.   The simplified rule very seldom reduces the amount of calculations.

   - If the structure is very simple, it is as fast, or even faster, to calculate the effects of actions in case of fire as to calculate the reduction factor η_{fi} and then multiply the effects of actions at room temperature by this factor.
   - If the structure is complex, it is nowadays a common practice to determine the effects of actions by means of a numerical tool, say a finite element or a displacement based program. In this case, it is a small task to analyse the structure for a couple of additional load combinations, namely the load combinations in case of fire. The most important task is the discretisation of the structure. If, on the other hand, it is decided to apply the simplified rule, it is not always a simple task to determine which one from the load combinations at room temperature is the fundamental one. Equation 2.13 has thus to be applied several times with different actions considered as the dominant load condition.

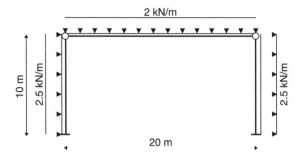

2 kN/m

10 m

2.5 kN/m

2.5 kN/m

20 m

Fig. 2.2 Characteristic loads on a portal frame

2. The simplified rule can lead to results that are not statically correct. Let us consider an element in which the permanent load induces an axial force and the variable load, for example the wind, induces a bending moment. Any load combination at room temperature will induce effects of action in the element that will be of the following type:

$$E_d = (\gamma_G N_k \ ; \ \gamma_Q M_k) \quad \text{for example } E_d = (1.35 \ N_k \ ; \ 1.50 \ M_k)$$

The effects of actions in case of fire have, according to the general Equation 2.5, the following expression:

$$E_{fi,d} = (N_k \ ; \ \psi_{1,1} M_k) \quad \text{for example } E_{fi,d} = (N_k \ ; \ 0.20 \ M_k)$$

It is obvious that any multiplication of E_d by a scalar factor will lead to a result which is different from $E_{fi,d}$.

As an example, let us consider the simple plane frame shown on Figure 2.2 submitted to a permanent vertical load of 2 kN/m on the beam and a wind load of 2.5 kN/m on the columns. The values indicated for the loads are characteristic values. The dead weight of the columns is supposed to be negligible.

The characteristic values of the effects of actions at the base of the columns are easily calculated.

$$(N_k; M_k) = (2 \, \text{kN/m} \times 20 \, \text{m}/2; 2.5 \, \text{kN/m} \times 10 \, \text{m} \times 5 \, \text{m})$$
$$= (20 \, \text{kN}; 125 \, \text{kNm})$$

The design values of the effects of action are determined for the design at ambient temperature.

$$(N_d; M_d) = (1.35 \times 20 \, \text{kN}; 1.50 \times 125 \, \text{kNm})$$
$$= (27 \, \text{kN}; 187.5 \, \text{kNm})$$

Application of Equation 2.5 directly yields the effects of actions in case of fire as

$$(N_{fi,d}; M_{fi,d}) = (1.00 \times 20 \, \text{kN}; 0.20 \times 125 \, \text{kNm})$$
$$= (20 \, \text{kN}; 25 \, \text{kNm})$$

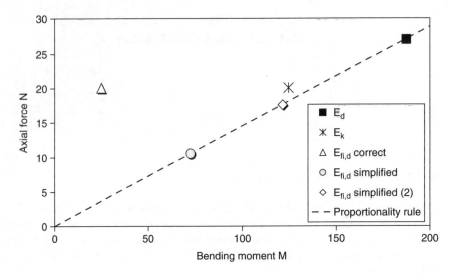

Fig. 2.3 Different effects of actions

However, the application of "simplified" rule leads to:

$$\eta_{fi} = (1.00 \times 40\,\text{kN} + 0.2 \times 50\,\text{kN})/(1.35 \times 40\,\text{kN} + 1.5 \times 50\,\text{kN})$$
$$= 50/129 = 0.388$$
$$(N_{fi,d}; M_{fi,d}) = (0.388 \times 27\,\text{kN}; 0.388 \times 187.5\,\text{kNm})$$
$$= (10.5\,\text{kN}; 72.7\,\text{kNm})$$

It can be seen that the application of the simplified rule requires more effort than the application of the exact rule and that it yields an axial force that is only half of the correct value, whereas the bending moment is almost three times the correct value!

This comparison is illustrated in Figure 2.3.

It can also be calculated that the eccentricity of the load in the correct evaluation is 25 kNm/20 kN = 1.25 m, whereas the value derived from the simplified rule is calculated as 72.7 kNm/10.5 kN = 6.92 m. The consequence might not be negligible, especially in members that have a highly non symmetrical M-N interaction diagram, for example in reinforced concrete, prestressed concrete or composite steel-concrete members.

As a further simplification to the already simplified rule, an arbitrary value of $\eta_{fi} = 0.65$ may be used, in which case it is indeed trivial to determine the effects of actions in the fire situation. Whether this approach can still be considered as a performance-based design is highly questionable. The point corresponding to this further simplification is noted as "$E_{fi,d}$ simplified(2)" on Figure 2.3.

From here on in this book, the basic rule according to Equation 2.5 will systematically be used.

Chapter 3

Thermal Action

3.1 Fundamental principles

The calculation approaches to model the thermal action produced by a fire on a structure is described in Eurocode 1 Part 2. Different representations of the effects of fire are given: temperature–time relationships, zone models or localised models. Some considerations are given hereafter on various aspects related to the calculation of temperatures in steel sections depending of the type of representation of the fire.

The fire severity to be used for design depends on the legislative environment and on the design philosophy. In a *prescriptive* code, the design fire severity is usually prescribed by the code with little or no room for discussion. In a *performance based* code, the design fire is usually recommended to be a complete burnout, or in some cases a shorter time of fire exposure which only allows for escape, rescue, or firefighting (Buchanan 2001). The *equivalent time* of a complete burnout is the time of exposure to the standard test fire that would result in an equivalent impact on the structure.

3.1.1 Eurocode temperature-time relationships

In case of a fully developed fire, the action of the fire is most often represented by a temperature–time curve, i.e. an equation describing the evolution with time of the unique temperature that is supposed to represent the environment in which the structure is located.

3.1.1.1 Nominal fire curves

This equation can be one of the nominal curves given in the Eurocode. They are:

1. The standard curve (sometimes called the standard ISO 834 curve, given in prEN13501-2).

$$\theta_g = 20 + 345\log_{10}(8t + 1) \tag{3.1}$$

   This curve is used as a model for representing a fully developed fire in a compartment.
2. The external fire curve.

$$\theta_g = 20 + 660(1 - 0.686\,e^{-0.32t} - 0.313\,e^{-3.8t}) \tag{3.2}$$

This curve is intended for the outside of separating external walls which exposed to the external plume of a fire coming either from the inside of the respective fire compartment, from a compartment situated below or adjacent to the respective external wall. This curve is **not** to be used for the design of external steel structures for which a specific model exists, see section 4.4.

3.   The hydrocarbon curve.

$$\theta_g = 20 + 1080(1 - 0.325\,e^{-0.167t} - 0.675\,e^{-2.5t}) \tag{3.3}$$

This curve is used for representing effects of a hydrocarbon fire.

In these equations,

θ_g is the gas temperature in the compartment (Eq. 3.1 and 3.3) or near the steel member (Eq. 3.2), in °C,

t is the time, in minutes.

These nominal fire curves are illustrated on Figure 3.1.

3.1.1.2   Equivalent time

Annex F of Eurocode 1 contains a method yielding an equivalent time of fire exposure that brings back the user to the utilisation of the standard temperature–time curve. The method is based on three parameters representing three physical quantities, namely the design fire load, the quantity and types of openings and the thermal properties of the walls. A fairly simple equation gives, as a function of these three parameters the duration of the standard fire that would have the same effect on the structure than a real fire that could occur in the relevant conditions. Equivalent time methods

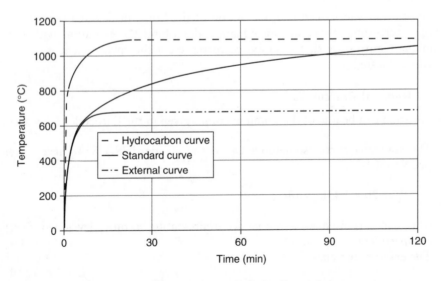

Fig. 3.1 Three different nominal fire curves as specified in Eurocodes

are nowadays considered somewhat out-dated and other more refined models exist that allow representing the influence of the conditions on the severity of a real fire, see below.

3.1.1.3 Parametric temperature–time curves

The temperature representing the fire can also be given by the parametric temperature–time curve model from annex A of Eurocode 1. This annex presents all equations required to calculate the temperature–time curve based on the value of the parameters that describe the particular situation. The model is valid for fire compartments up to 500 m² of floor area, maximum height of 4 meters without openings in the roof. The same equations are presented here in a somewhat more logical way, i.e. in the order in which they have to be used.

The input data are:

- Thermal properties of the walls, ceiling and floor, namely thermal conductivity λ in W/mK, specific heat c in J/kgK and density ρ in kg/m³.
- Geometric quantities such as total area of walls, ceiling and floor including openings A_t in m², total area of vertical openings of all walls A_v in m² and weighted average of opening heights on all walls h_{eq} in m.
- Design value of the fire load density $q_{t,d}$ in MJ/m², related to A_t.
- Fire growth rate, slow, medium or high.

The successive steps are:

1. Evaluate the wall factor b according to Equation 3.4 (somewhat more complex equations are given for enclosure surfaces made of several layers of different materials or enclosure surfaces made of different materials).

$$b = \sqrt{c\rho\lambda} \tag{3.4}$$

2. Evaluate the opening factor O according to Equation 3.5.

$$O = A_v\sqrt{h_{eq}}/A_t \tag{3.5}$$

where h_{eq} is the weighted average of the window heights $\left(\sum A_{vi}h_i/\sum A_{vi}\right)$

3. Evaluate the factor Γ according to Equation 3.6 (Γ factors higher than 1 will yield a heating phase of the parametric curve that is hotter than the ISO curve and vice versa).

$$\Gamma = \left(\frac{O/0.04}{b/1160}\right)^2 \tag{3.6}$$

4. Determine the shortest possible duration of the heating phase t_{lim} in hours, depending on the fire growth rate. $t_{lim} = 25$ minutes, i.e. 5/12 hour, for slow fire growth rate; $t_{lim} = 20$ minutes, i.e. 1/3 hour, for medium fire growth rate; $t_{lim} = 15$ minutes, i.e. ¼ hour, for fast fire growth rate.

5. Evaluate the duration of the heating phase t_{max} in hours according to Equation 3.7

$$t_{max} = 0.2 \times 10^{-3} q_{t,d}/O \tag{3.7}$$

6. If $t_{max} > t_{lim}$ then the fire is ventilation controlled and;
    - The temperature during the heating phase, i.e. until $t = t_{max}$, is given by Equation 3.8.

$$\theta_g = 20 + 1325(1 - 0.324\, e^{-0.2t^*} - 0.204\, e^{-1.7t^*} - 0.472\, e^{-19t^*}) \tag{3.8}$$

with $t^* = \Gamma t$
    - The temperature during the cooling down phase is given by Equation 3.9.

$$\theta_g = \theta_{max} - 625(t^* - t^*_{max}) \qquad \text{for } t^*_{max} \le 0.5 \tag{3.9a}$$
$$\theta_g = \theta_{max} - 250(3 - t^*_{max})(t^* - t^*_{max}) \quad \text{for } 0.5 < t^*_{max} < 2.0 \tag{3.9b}$$
$$\theta_g = \theta_{max} - 250(t^* - t^*_{max}) \qquad \text{for } 2.0 < t^*_{max} \tag{3.9c}$$

where
$t^*_{max} = \Gamma t_{max}$
θ_{max} is given by Equation 3.8 in which $t^* = t^*_{max}$

7. If $t_{max} \le t_{lim}$ then the fire is fuel controlled and the steps are:
    - Evaluate the modified opening factor O_{lim} according to Equation 3.10.

$$O_{lim} = 0.1 \times 10^{-3}\, q_{t,d}/t_{lim} \tag{3.10}$$

    - Evaluate the modified factor Γ according to Equation 3.11.

$$\Gamma_{lim} = \left(\frac{O_{lim}/0.04}{b/1160}\right)^2 \tag{3.11}$$

    - If $O > 0.04$ and $q_{t,d} < 75$ and $b < 1160$, Γ_{lim} obtained from Equation 3.11 has to be multiplied the factor k given by Equation 3.12.

$$k = 1 + \left(\frac{O - 0.04}{0.04}\right)\left(\frac{q_{t,d} - 75}{75}\right)\left(\frac{1160 - b}{1160}\right) \tag{3.12}$$

    - The temperature during the heating phase, i.e. until $t = t_{lim}$, is given by Equation 3.8 in which $t^* = \Gamma t_{lim}$
    - the temperature during the cooling down phase is given by Equation 3.13.

$$\theta_g = \theta_{max} - 625(t^* - t^*_{lim}) \qquad \text{for } t^*_{max} \le 0.5 \tag{3.13a}$$
$$\theta_g = \theta_{max} - 250(3 - t^*_{max})(t^* - t^*_{lim}) \quad \text{for } 0.5 < t^*_{max} < 2.0 \tag{3.13b}$$
$$\theta_g = \theta_{max} - 250(t^* - t^*_{lim}) \qquad \text{for } 2.0 < t^*_{max} \tag{3.13c}$$

where $t^*_{lim} = \Gamma_{lim} t$

This parametric fire model described in Annex A of EN 1991-1-2 has its root in the parametric model that was present in ENV 1991-1-2, this one based on work done in Sweden by Petersson, Thelandersson and Magnusson and later reformulated by Wickström. In addition, Franssen (2000) has introduced three modifications to the ENV model, namely:

1. The first one deals with the equations that allow calculating the wall factor b for walls with several layers made of different materials, as is the case, for example, for a brick wall covered by a layer of plaster. The equation has been improved in order to take into account in a more precise manner the amount of energy introduced in layered walls when they are heated.
2. The concept of a minimum duration of the heating phase, t_{lim}, has been introduced that marks the transition from a fuel to an air control situation. The idea is that, no matter how big the openings are and no matter how small the fire load is, it nevertheless requires a certain amount of time, namely t_{lim}, to burn any piece of furniture. A typical example could be the burning of one single car in an open car park.
3. The factor k defined by Equation 3.12 has been introduced to take into account the effects of large openings that, in a fuel controlled situation, vent the compartment and, thus, limit the rise in temperatures. This coefficient has been calibrated in order to obtain the best fit between the model and a set of 48 experimental fire tests.

In the background document of EN 1991-1-2, ProfilARBED 2001, two additional modifications have been introduced. Firstly, the coefficient of Equations 3.7 and 3.10 have been given different values, namely 0.001 and 0.002 whereas a single value of 0.0013 had been proposed by Franssen (2000). Secondly, the notion that t_{lim} depends on the fire growth rate has been introduced. These modifications were introduced because they improved the fit between the model and the set of available experimental results. The first one has yet the consequence that the model is no longer continuous; a hiatus now exists between the temperature–time curves in the fuel controlled regime and those in the air controlled regime.

3.1.1.4   Zone models

One-zone models and two-zone models also produce as a result a temperature that represents the temperature of the gas in the compartment, for the one-zone models, or in each zone, for the two zone models. These models are based on the hypothesis that the situation with regard to temperature is uniform in the compartment (one-zone models) or in each of the upper and the lower zone that exist in the compartment (two-zone) models. Physical quantities such as the properties of the walls and the openings are not lumped into one single parameter as is the case for the parametric fire models but, on the contrary, each opening can be represented individually and each wall can be represented with its own thermal properties. The temperature evolution is not prescribed by a predetermined equation as in the parametric fire models but results from the integration on time of the differential equations expressing the equilibrium of mass and of energy in the zones. Application of such models thus requires the utilisation of numerical computer software, see for example the software OZone developed at the university of Liege (Cadorin and Franssen 2003, and Cadorin et al. 2003).

3.1.1.5 *Heat exchange coefficients*

For all these situations when the action of the fire around the structural member is represented by a unique temperature, Eurocode 1 gives the equation that has to be used in order to calculate at any time t the net heat flux reaching a steel member. Taking into account the fact that the emissivity of the fire and the configuration factor may be taken as equal to 1, and the fact that the radiation temperature may be taken as the gas temperature, the net heat flux can be calculated by Equation 3.14. It shows that the net flux is made of a convection term and a radiation term.

$$\dot{h}_{net} = \alpha_c(\theta_{g,t} - \theta_{m,t}) + \varepsilon_m \sigma(\theta_{g,t}^4 - \theta_{m,t}^4) \tag{3.14}$$

where:

α_c is the coefficient of heat transfer by convection,
$\theta_{g,t}$ is the temperature of the gas around the member (in K),
$\theta_{m,t}$ is the surface temperature of the member (in K),
ε_m is the surface emissivity of the member,
σ is the Stephan Boltzmann constant $(=5.67\,10^{-8}\,\mathrm{W/m^2 K^4})$.

The surface emissivity is taken as 0.7 for steel carbon steel, 0.4 for stainless steel and 0.8 for other materials for which related design parts of EN 1992 to 1996 and 1999 give no specific value, e.g. concrete.

The value to be used for the coefficient of convection (α_c) depends on the fire curve that is considered and on the surface conditions, either on a surface exposed to the fire or on the unexposed side, for example the top surface of a concrete slab heated from underneath. Table 3.1 gives the recommended values of α_c for different surface conditions.

3.1.2 *Eurocode localised fire, flame not impacting the ceiling*

In case of a localised fire, the flame length L_f in meters is calculated according to Equation 3.15 known in the literature as the Heskestad flame height correlation, Heskestad 1983.

$$L_f = 0.0148\,Q^{0.4} - 1.02\,D \tag{3.15}$$

Table 3.1 Coefficient of convection for different surface conditions

	α_c (W/m²K)
Unexposed side of separating elements	
Possibility 1: radiation considered separately	4
Possibility 2: radiation implicitly contained	9
Surface exposed to the fire	
Standard curve or external fire curve	25
Hydrocarbon curve	50
Parametric fire, zone fire models or external members	35

where:

Q is the rate of heat release of the fire [W],
D is the characteristic length of the fire, its diameter for example for a circular shape [m].

It has to be noted that, for low values of the heat release and large values of the characteristic length, this equation can yield negative flame heights. This is of course not physically possible. It simply indicates that a single flaming area of diameter D has broken down into several smaller separate zones and the equation should be applied individually for each zone (Heskestad 1995).

If the flame length from the fire source does not reach the ceiling, only the temperature evolution along the flame length is given in Eurocode 1.

The position of the virtual origin, z_0 in meters, is first calculated according to Equation 3.16. The values of z_0 are negative because the virtual origin is lower than the fire source.

$$z_0 = 0.00524Q^{0.4} - 1.02D \qquad (3.16)$$

The evolution of the temperature in the plume in °C along the vertical axis of the flame is given by Equation 3.17.

$$\theta = 20 + 0.25Q_c^{2/3}(z - z_0)^{-5/3} \leq 900 \qquad (3.17)$$

where Q_c is the convective part of the rate of heat release Q,
with $Q_c = 0.8Q$ by default.

It is then the task of the user to make his own hypotheses to calculate the heat flux emitted by the flame that reaches the surface of the structural member. In order to estimate this flux, hypotheses have to be made on the shape of the flame, for example cylindrical, and on the temperature distribution in the horizontal plane of the flame, for example constant temperature. The view factors from different surfaces of the flame to the member can then be evaluated according to Annex G of Eurocode 1 and the flux finally calculated. Although this is not explicitly stated in the Eurocode, Annex B on thermal actions for external members could be of good guidance here, for example for estimating the emissivity of the flame.

Figure 3.2 shows the flame length (L_f) as a function of the diameter (D) of a cylindrical fire source for different rate of heat release densities RHR_f. It shows that for densities of 250 kW/m², recommended for example in dwellings, hospital and hotel rooms, offices, classrooms, but also shopping centres and public spaces in transport buildings, the flame height never exceeds 2.07 meters (obtained for $D = 8$ m) and thus never reaches the ceiling. The temperature in the plume at the level of the ceiling will thus always be lower than the temperature at the tip of the flame, i.e. lower than 520°C. This is clearly not realistic and this model should not be applied for such a low heat release rate density.

For densities of 500 kW/m² recommended in libraries and theatres (cinema), longer flame length are calculated and the flame may eventually reach the ceiling, in which case the model described below in Section 3.1.3 can be applied.

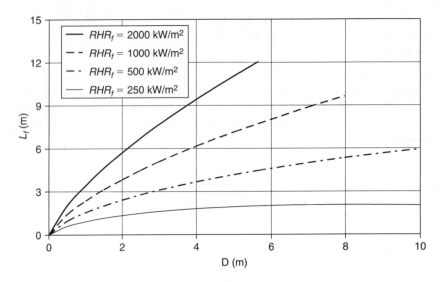

Fig. 3.2 Flame length as a function of the diameter of the fire

Note: The curves in Figure 3.2 are not continued beyond the limits of application mentioned in Eurocode 1 for the models specified in Sections 3.1.2 and 3.1.3, i.e. $D = 10\,m$ and $Q = 50\,MW$.

In fact, as long as the fire is not severe enough to produce flames that impact the ceiling, the threat posed by the fire to the structure supporting the ceiling is not very severe. It can indeed be calculated that with this model the temperature at the tip of the flame is always equal to 520°C. Consequently, the evaluation of the effect of the local fire on the structure of the ceiling at this stage is usually not performed and this effect is neglected.

Cases when consideration should be given to the situation of a localised fire not impacting the ceiling comprise, for example, the heat flux from a fire to a column that is engulfed in the fire in a compartment with a high floor to ceiling distance. It can be assumed that the column is located at the centre line of the flame and Equation 3.17 can then be used to estimate the temperature of the flame surrounding the column.

Figure 3.3 shows the evolution of the temperature in the centreline of the flame as a function of the height, for different diameters of the cylindrical fire source, considering a rate of heat release densities RHR_f of 500 kW/m² The horizontal line at the level of 520°C corresponds to the tip of the flame.

3.1.3 *Eurocode localised fire, flame impacting the ceiling*

In case of a localised fire with the flame tip impinging on the ceiling, the total heat flux received by the structure at the level of the ceiling is given as a function of geometrical parameters and of the size and rate of heat release of the fire.

The model given in the Eurocode is based on experimental tests made in Japan (Hasemi et al. 1984, Ptchelintsev et al. 1995, Hasemi et al. 1995,

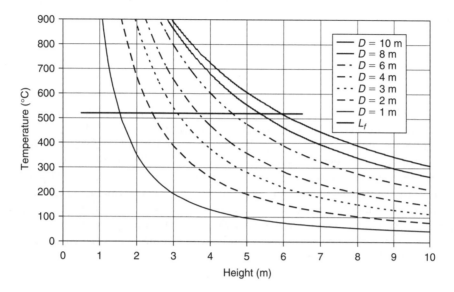

Fig. 3.3 Evolution of the temperature for $RHR_f = 500\,kW/m^2$

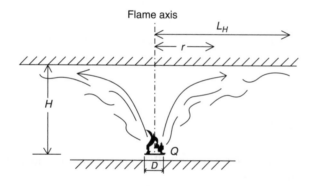

Fig. 3.4 Localised fire, flame impacting the ceiling

Wakamatsu et al. 1996). These tests were small scale tests, made in steady state conditions. The flame from a gas burner was impinging a panel placed above the burner. The air flow was undisturbed in the sense that no vertical panels were placed laterally to simulate the effect of eventual compartment walls, Figure 3.4.

The original equations of Hasemi were slightly modified within the European research "Development of design rules for steel structures subjected to natural fires in large compartments" (Schleich et al. 1999). It has been demonstrated (Franssen et al. 1998, and Schleich et al. 1999) that this model yields acceptable results in transient situations within real compartments for heat release rates up to 50 MW. Experimental validations of Hasemi's model on full scale tests can also be found in Wakamatsu et al. (2002).

The non-dimensional square root of the Froude number Q_D^* is calculated according to Equation 3.18.

$$Q_D^* = \frac{Q}{1.11 \times 10^6 D^{2.5}} \tag{3.18}$$

This number is big for fires in which the velocity of the gas is high compared to the effects of buoyancy such as in jet flames or gas burner. If the velocity of the premixed gas in the burner is high enough, the length of the flame is virtually independent from the direction of the flame relative to the direction of gravity. This number is small for fires in which the velocity is low compared to the effects of buoyancy such as in pool fires.

The vertical position of the virtual source z' in meters is calculated according to Equation 3.19, first proposed by Hasemi and Togunaga, 1984. On the contrary to Equation 3.16 used to calculate basically the same physical quantity, Equation 3.19 has been arranged in such a way as to yield positive values when the virtual source is located under the fire source.

$$\begin{aligned} z' &= 2.4D(Q_D^{*2/5} - Q_D^{*2/3}) \quad \textit{when } Q_D^* < 1.0 \\ z' &= 2.4\,D(1.0 - Q_D^{*2/5}) \quad \textit{when } Q_D^* \geq 1.0 \end{aligned} \tag{3.19}$$

Another non-dimensional square root of the Froude number, Q_H^*, is calculated on the basis of the vertical distance H between the fire source and the ceiling according to Equation 3.20.

$$Q_H^* = \frac{Q}{1.11 \times 10^6 H^{2.5}} \tag{3.20}$$

The length of the flame, $H + L_h$, is calculated according to Equation 3.21, with L_h as the horizontal flame length.

$$H + L_h = 2.9H(Q_H^*)^{0.33} \tag{3.21}$$

The non dimensional ratio y is calculated according to Equation 3.22. This is the ratio between the distance from the virtual source to the point along the ceiling where the flux is calculated, on one hand, and the distance from the virtual source and the tip of the flame, on the other hand.

$$y = \frac{z' + H + r}{z' + H + L_h} \tag{3.22}$$

where r is the horizontal distance from the vertical axis of the fire to the point under the ceiling where the flux is calculated.

The heat flux received by the structure at the level of the ceiling, $\dot{h}$ in W/m², is given by Equation 3.23.

$$\begin{aligned}
\dot{h} &= 100\,000 & when\ y &\leq 0.30 \\
\dot{h} &= 136\,300 - 121\,000\,y & when\ 0.30 &< y < 1.0 \\
\dot{h} &= 15\,000\,y^{-3.7} & when\ 1.0 &\leq y
\end{aligned} \tag{3.23}$$

The net heat flux is the difference between the flux received by the member and the heat energy lost by the member to the environment by convection and radiation, see Equation 3.24

$$\dot{h}_{net} = \dot{h} - \alpha_c(\theta_{m,t} - 293) - \varepsilon_m\sigma(\theta_{m,t}^4 - 293^4) \tag{3.24}$$

In case of several separate localised fire sources, each source is supposed to generate a heat flux calculated according to Equation 3.23 and the contributions are added, but the sum of all contributions should not exceed 100 kW/m². No experimental evidence can justify this approximation, but it is believed to be on the conservative side because, in reality, the ceiling jets from each fire source cannot be added and may even counteract each other. For a structural element that is located between two sources, for example, it may happen that the mass flows coming from each source more or less annihilate each other.

Although this is not physically correct, this model is sometimes used in order to evaluate the heat flux from a localised fire to horizontal beams that are not located directly under the ceiling but at some distance below the ceiling. This is the case, for example, for the lower members of a steel truss supporting the ceiling. The model is used simply replacing the vertical distance between the fire source and the ceiling by the vertical distance between the fire source and the member. It is generally accepted that this utilisation of the model yields results that are on the conservative side.

Because only the heat flux at the level of the ceiling is given, this model cannot be used as such for evaluating the effect of the fire on a column. Users must rely on the literature for evaluating the effects on the columns of a localised fire that impacts the ceiling, see for example Kamikawa et al. 2001.

3.1.4 CFD models in the Eurocode

Eurocode 1 allows the utilisation of CFD (Computational Fluid Dynamics) models. Although prEN 1991-1-2 states under Clause 3.3.2 (2) that a method is given in Annex D for the calculation of thermal actions in case of CFD models, this annex simply gives general principles that form the base of the method and must be respected when establishing a software that allows application of this method in order to estimate the temperature field in the compartment. No guidance is provided on the manner to deduce the heat flux on the surface of the structural elements from the temperatures calculated in the compartment by the CFD model. In fact, this topic is still nowadays a subject of ongoing research activities and is probably premature to layout recommendations in a Code.

The Eurocode opens the door for application of the CFD models in fire safety engineering but, at the moment, this is not yet standard practice and can be made only by very experienced user. This is probably why the Eurocode has foreseen that the

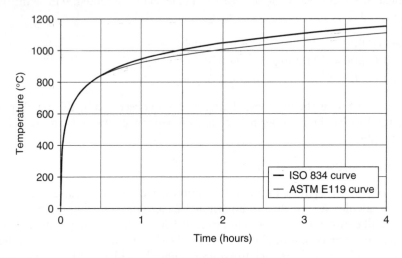

Fig. 3.5 Comparison between ISO 834 and ASTM E119 fire curves

National Annex in each country may specify the procedure for calculating the heating conditions from advanced fire models.

3.1.5 North American time–temperature relationships

North American standards rely on large scale fire resistance tests to assess the fire performance of building materials and structural elements. The time temperature curve used in fire resistance tests is called the *standard fire*. Full size tests are preferred over small scale tests because they allow the method of construction to be assessed, including the effects of thermal expansion, shrinkage, local damage and deformation under load.

In US, standard fire resistance tests are carried out as per the specifications in ASTM E119 (ASTM 2002), NFPA 251 (NFPA 1999), UL 263 (UL 2003). The standard time temperature curves from ASTM E119 and ISO 834 are compared in Figure 3.5. They are seen to be rather similar. All other international fire resistance test standards specify similar time temperature curves.

The development of a standard for characterizing fire exposure scenarios is currently underway in United States. SFPE has recently published an Engineering guide (SFPE 2004) which can be used to develop design fire scenarios for various compartment types.

Other national standards include British Standard BS 476 Parts 20-23 (BSI 1987), Canadian Standard CAN/ULC-S101-04 (ULC 2004) and Australian Standard AS 1530 Part 4 (SAA 1990).

The ASTM E119 curve is defined by a number of discrete points, which are shown in Table 3.2, along with the corresponding ISO 834 temperatures. Several equations approximating the ASTM E119 curve are given by Lie (1995), the simplest of which gives the temperature T (°C) as

$$T = 750[1 - e^{(-3.79553\sqrt{t_h})}] + 170.41\sqrt{t_h} + T_0 \qquad (3.25)$$

where t_h is the time (hours).

Table 3.2 ASTM E119 and ISO 834 time temperature curves

Time (min)	ASTM E119 Temperature (°C)	ISO 834 Temperature (°C)
0	20	20
5	538	576
10	704	678
30	843	842
60	927	945
120	1010	1049
240	1093	1153
480	1260	1257

3.2 Specific considerations

3.2.1 Heat flux to protected steelwork

The procedures that have to be used to calculate the temperature in the steel members will be given in Chapter 4. For undertaking such temperature calculations, one particular question that has to be addressed is the nature of boundary conditions, especially for protected sections.

For unprotected sections indeed, the heat flux introduced in the section is present in the equation used for evaluating the temperature in the section, see Equation 4.1. This flux is easily calculated according to Equation 3.14 if the fire is represented by a temperature-time curve. In the case of a localised fire impinging the ceiling (see Section 3.1.3), the effect of the fire is directly given as a net heat flux (see Equation 3.24), and this can also be used directly for calculating the steel temperature. For localised fires not impinging the ceiling (see Section 3.1.2), it is also possible to calculate the net heat flux, for example from Equation 3.14 if the fire is surrounding a column and the temperature of the hot gases is given by Equation 3.17.

For protected steelworks on the other hand, the equation used for calculating the temperature is based on the gas temperature, see Equation 4.7. This equation can be used as such in the cases where the fire is represented by a gas temperature, see sections 3.1.1 and 3.1.2. Equation 4.7 cannot yet be applied directly if the effect of the fire is given as an impinging flux, see section 3.1.3. A procedure has to be established to transform the impinging heat flux into an equivalent gas temperature.

This procedure is based on the assumption (also made in Eq. 4.7) that the surface temperature of the protection, $\theta_{m,t}$, is equal to the gas temperature, $\theta_{g,t}$.

If the fire is represented by a temperature–time curve and the boundary condition is taken as expressed by Equation 3.14, then net heat flux is obviously equal to 0 when $\theta_{m,t} = \theta_{g,t}$. If the same condition $\dot{h}_{net} = 0$ is imposed in Equation 3.24, it is then possible to derive the equivalent temperature of the gas leading to the situation when the flux received by the surface is exactly equal to the flux reemitted by that surface. This equivalent temperature of the gas is given by the solution of Equation 3.26.

$$\dot{h} = \alpha_c(\theta_{g,t} - 293) + \varepsilon_m\sigma(\theta_{g,t}^4 - 293^4) \tag{3.26}$$

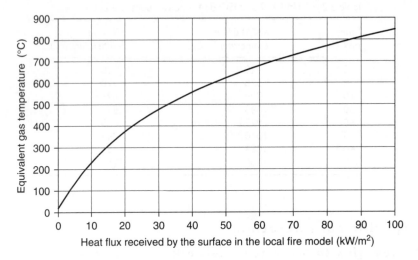

Fig. 3.6 Evolution of the equivalent gas temperature as a function of the impinging flux

Figure 3.6 shows the relation between the equivalent gas temperature and the heat flux given by the local model if the values $\alpha_c = 35$ W/m^{2}K and $\varepsilon_m = 0.8$ are introduced in Equation 3.26. It has to be underlined that, because the impinging heat flux given by the local model has a maximum value of 100 kW/m^2, whatever the geometrical conditions and whatever the power released by the fire, the equivalent gas temperature cannot be higher than 847°C and, hence, the steel temperature of a steel member heated by this local model cannot reach temperatures higher than 847°C.

The structure of Equation 3.26 also shows that higher values of coefficients of heat transfer, α_c and ε_m, lead to lower values of the steel temperature because, for a given impinging flux, the reemitted flux will be greater!

3.2.2 Combining different models

When the fire is localised, a 2-zone model will give as a result, amongst other things, the evolution of the temperature in the hot zone. This temperature must be regarded as the average temperature that one could observe or measure in the hot zone during a test or a real fire. In addition to that, it must be considered that a more severe thermal attack is imposed on structural elements that are located in the near vicinity of the fire source. For example, if one car is burning in a car park, stratification will indeed be observed and a 2-zone model allows predicting the temperature in the hot zone, at least at a reasonable distance from the burning car. Just above the car, the situation is certainly very different, with a much severe thermal attack by direct radiation from the flames produced by the burning car. This effect is evaluated by the localised fire model described in Annex C of EN 1991-1-2, see for example Section 3.1.3 here above.

The Eurocode recommends that the combination of the results obtained by the 2-zone model and by the localised fire model may be considered, in order to get a more accurate temperature distribution along the members. According to the Eurocode, the combination simply considers taking at each location and at any time the maximum of the effect given by the two models. It is the opinion of the authors of this book that the combination should generally be made. If, for example, the fire source is located under a simply supported truss girder at, say, ¼ of the span, it is not possible to predetermine which member of the truss will fail first, either a bar just above the fire source, possibly loaded to a lower level but submitted to a more severe attack from the fire, or a bar at mid span of the truss, presumably loaded to a higher level but heated by the average temperature of the hot zone, i.e. by a less severe thermal attack.

When flash-over occurs and the situation in the compartment turns into a one-zone situation, there is no localised fire source anymore and, thus, no need to combine. As a consequence, it may happen that a structural member located in the vicinity of the source is submitted to a less severe thermal attack immediately after the flash-over than immediately before the flash-over. Usually, because the flash-over is accompanied by a sudden increase in the rate of heat release of the fire, the duration of this apparent bias is very short and its effect can hardly be seen on the temperature curve in the steel member.

3.3 Examples

3.3.1 Localised fire

A car is burning with a heat release rate of 5 MW in a car park with a floor to ceiling distance of 2.80 m. What is the flux received by a steel element located under the ceiling at a horizontal distance of 5 meters from the centre of the car?

Characteristic length of the fire source
The fire source is assumed to be rectangular with a surface area equal to $10 \, m^2$ (2×5), which means an equivalent diameter D of 3.57 m.

Length of the flame
$L_f = 0.0148 \, (5 \, 10^6)^{0.4} - 1.02 \times 3.57 = 3.44 \, m$
The flame is impinging the ceiling.

Non dimensional Froude number
$Q_D^* = 5 \, 10^6 / (1.11 \, 10^6 \times 3.57^{2.5}) = 0.187$

Position of the virtual source
$z' = 2.4 \times 3.57(0.187^{0.4} - 0.187^{0.67}) = 1.58 \, m$

Non dimensional Froude number
$Q_H^* = 5 \, 10^6 / (1.11 \, 10^6 \times 2.30^{2.5}) = 0.561$

Note that the fire source is supposed to be 0.50 m above the ground, hence the vertical distance from the fire source to the ceiling, H, is equal to 2.30 m.

Length of the flame
$H + L_h = 2.9 \times 2.30 \times 0.561^{0.33} = 5.50 \, m$

Non dimensional ratio
$y = (1.58 + 2.30 + 5.00)/(1.58 + 5.50) = 8.88/7.08 = 1.25$

Impinging flux
$\dot{h} = 15\,000/1.25^{3.7} = 6491\,\text{W/m}^2$

According to Figure 3.5, the temperature of a steel member submitted to such a flux cannot reach a temperature higher than 167°C.

3.3.2   Parametric fire – ventilation controlled

A fire compartment is rectangular in plan of size of 3 m by 6 m. The floor to ceiling distance is 2.5 m. The design fire load is 750 MJ/m² and the rate of fire growth in the compartment is "slow". The compartment partitions are made of normal weight concrete, $C = 1100\,\text{J/kgK}$, $\rho = 2300\,\text{kg/m}^3$, $\lambda = 1.2\,\text{W/mK}$. One window, 2 meters wide and 1 meter high, and one door, 1 meter wide and 2.1 meters high, open in the walls. Calculate the parametric fire curve.

Wall factor b
$b = (1100\,\text{J/kgK} \times 2300\,\text{kg/m}^3 \times 1.2\,\text{W/mK})^{0.5} = 1742\,\text{J/m}^2\text{s}^{0.5}\text{K}$

Total area of vertical openings
$A_v = 2\,\text{m} \times 1\,\text{m} + 1\,\text{m} \times 2.1\,\text{m} = 4.1\,\text{m}^2$

Total area of enclosures
$A_t = 2\,(3\,\text{m} \times 6\,\text{m} + 3\,\text{m} \times 2.5\,\text{m} + 6\,\text{m} \times 2.5\,\text{m}) = 81\,\text{m}^2$

Weighted average of window heights
$h_{eq} = (2\,\text{m}^2 \times 1\,\text{m} + 2.1\,\text{m}^2 \times 2.1\,\text{m})/4.1\,\text{m}^2 = 1.56\,\text{m}$

Opening factor
$O = 4.1 \times 1.56^{0.5}/81 = 0.0633\,\text{m}^{0.5}$

Factor Γ
$\Gamma = (0.0633/0.04)^2/(1742/1160)^2 = 1.11$

Fire load density
$q_{t,d} = 750 \times 18/81 = 167\,\text{MJ/m}^2$

Shortest possible duration of the heating phase
$t_{\lim} = 5/12 = 0.417\,\text{h}$ (25 min.)

Duration of the heating phase
$t_{max} = 0.2\,10^{-3} \times 167/0.0633 = 0.528\,\text{h}$ (31 min. 41 s)
The fire is ventilation controlled because $t_{\lim} < t_{max}$

Temperature during the heating phase can be computed using Equation 3.8. For example;

- At $t = 30$ min., i.e. $t = 0.5\,\text{h}$, $t^* = 1.11 \times 0.5 = 0.555$, $\theta_g = 856°\text{C}$.
- At $t = t_{max}$, $t^*_{max} = 1.11 \times 0.528 = 0.586$, $\theta_{max} = 863°\text{C}$.

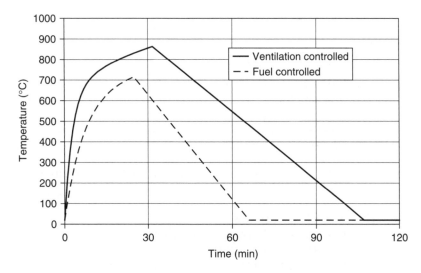

Fig. 3.7 Temperature–time curve according to the parametric model

Temperature during the cooling phase can be computed using Equation 3.9b. For example;

- At $t = 1.0$ h, $t^* = 1.11$, $\theta_g = 863 - 250(3 - 0.586)(1.11 - 0.586) = 547°$C.
- At $t = 1.787$ h, $t^* = 1.984$, $\theta_g = 863 - 250(3 - 0.586)(1.984 - 0.586) = 20°$C.

The complete temperature–time is plotted as a continuous line on Figure 3.7

3.3.3 *Parametric fire – fuel controlled*

How does the temperature–time curve get modified if the width of the window is increased to 3.4 meters?
   The parameters and factors that are modified are:

Total area of vertical openings
$A_v = 3.4\,\text{m} \times 1\,\text{m} + 1\,\text{m} \times 2.1\,\text{m} = 5.5\,\text{m}^2$

Weighted average of window heights
$h_{eq} = (3.4\,\text{m}^2 \times 1\,\text{m} + 2.1\,\text{m}^2 \times 2.1\,\text{m})/5.5\,\text{m}^2 = 1.42\,\text{m}$

Opening factor
$O = 5.5 \times 1.42^{0.5}/81 = 0.0809\,\text{m}^{0.5}$

Factor Γ
$\Gamma = (0.0809/0.04)^2/(1742/1160)^2 = 1.814$

Duration of the heating phase
$t_{max} = 0.2\ 10^{-3} \times 167/0.0809 = 0.413\,\text{h}$ (24 min. 46 s)
$t^*_{max} = 1.814 \times 0.413 = 0.749$
The fire is fuel controlled because $t_{max} \le t_{tlim}$

Modified opening factor
$O_{lim} = 0.1 \, 10^{-3} \times 167/0.417 = 0.04005$

Modified factor Γ
$\Gamma_{lim} = (0.04005/0.04)^2/(1\,742/1\,160)^2 = 0.444$

The temperature during the heating phase can be computed using Equation 3.8 in which $t^* = t\Gamma_{lim}$. For example;

- At $t = 20$ min., i.e. $t = 0.333$ h, $t^* = 0.444 \times 0.333 = 0.148$, $\theta_g = 680°C$.
- At $t = t_{lim}$, $t^* = 0.444 \times 0.417 = 0.185$, $\theta_{max} = 715°C$.

The temperature during the cooling phase can be computed using Equation 3.13b. For example;

- At $t = 1.0$ h, $t^* = 1.814 \times 1.0 = 1.814$,
  $\Theta_g = 715 - 250(3 - 0.749)(1.814 - 1.814 \times 0.417) = 120°C$.
- At $t = 1.1$ h, $t^* = 1.814 \times 1.10 = 1.991$
  $\Theta_g = 715 - 250(3 - 0.749)(1.991 - 1.814 \times 0.417) = 20°C$.

The complete temperature–time curve is now plotted on Figure 3.7 as a doted line.

Chapter 4

Temperature in Steel Sections

4.1 General

There are three main steps in the fire resistance analysis of steel structures. The first step is determining the fire temperature resulting from a given fire exposure scenario. This can be carried out using the methodologies described in Chapter 3. The second step is to establish the temperature history in the steel structure, resulting from fire temperature. The second step forms the basis for undertaking the structural analysis which is the third step of fire resistance calculation. The accuracy of calculations in the second stage is critical for obtaining realistic fire resistance predictions. In fact in many cases an estimate of fire resistance can be obtained based on predicted steel temperatures alone through the use of critical temperature failure criterion.

Establishing temperature history in the steel structure typically involves heat transfer analysis. For this simple or advanced calculation models described in later sections can be used.

The detailed steps for heat transfer calculations are contained in various manuals and guides. Eurocode provides different approaches for evaluating temperatures in steel section. The complexity of these methods, discussed later in the chapter, varies with the type of analysis and is related to the accuracy of temperatures. AISC steel design Manual (AISC 2005) does not contain the equations for such calculations. However, ASCE manual (ASCE 1992) details the procedure for temperature calculations under various exposure (2, 3 or 4 sided) types and boundary conditions.

One of the most important input for the heat analysis is the high temperature thermal properties of steel and insulation materials. The high temperature properties of steel as per North American and Eurocode practices are listed in Annex I. In addition, the room temperature thermal properties of common fire protection material are provided in Annex I. For simple calculation methods room temperature thermal properties may be sufficient to estimate the temperatures in steel structures.

4.2 Unprotected internal steelwork

4.2.1 Principles

If the temperature distribution in the cross section is supposed to be uniform, the temperature increase during a time increment is given by Equation 4.1

$$\Delta\theta_{a,t} = k_{sh}\frac{A_m/V}{c_a\rho_a}\dot{h}_{net}\,\Delta t \tag{4.1}$$

where:

$\Delta\theta_{a,t}$ is the steel temperature increase from time t to time $t + \Delta t$,
k_{sh} is the correction factor for the shadow effect, see below,
A_m is the surface area of the member per unit length,
V is the volume of the member per unit length,
c_a is the specific heat of steel,
ρ_a is the unit mass of steel,
$\dot{h}_{net}$ is the design value of the net heat flux per unit area,
Δt is the time interval.

Equation 4.1 is better understood if transformed into the form of Equation 4.2 which shows that it is just the expression of the conservation of energy between the quantity that penetrates in the section and the quantity used to modify the temperature and hence the enthalpy of the section.

$$\dot{h}_{net} k_{sh} A_m \, \Delta t = \Delta\theta_{s,t} c_a \rho_a V \tag{4.2}$$

In Equation 4.1, the ratio between the surface area of the member and the volume of the member, A_m/V, is the parameter characterising the cross section of the member that governs its heating. It is referred to in Eurocode 3 as the section factor. The higher the value of this factor, the thinner the section and, hence, the faster the heating of the section. Figure 4.1 hereafter, taken from Eurocode 3, shows how this parameter is calculated for different section configurations.

In fact, the term "*section factor*" is not meaningful because it contains no information about the physical characteristic that this factor represents. This parameter is sometimes referred to as "*the massivity factor*" which indicates at least what this factor is about but the problem nevertheless remains that this quantity is the highest for the most slender and less massive sections. In this text, the term "*thermal massivity*" will be introduced to designate the invert of the massivity factor. This thermal massivity indicates what physical effect it is related to, with the advantage that this quantity presents the highest values for the stockiest sections.

Table 4.1 indicates how each factor is related to the thickness t of a steel plate, either if this steel plate is used in an open section or if it is the wall of a steel tube.

The specific heat of steel c_a that is present in Equation 4.1 is given as a simple function of the steel temperature in EN 1993-1-2, see Equation I.1 in Annex I.

$\dot{h}_{net}$ is calculated as mentioned in Chapter 3.

k_{sh}, the correction factor for the shadow effect, stems from the fact that, at least in a furnace test, the steel section is mainly heated by the radiation that originates from the walls of the furnace and from the flames of the burners. In that case, there cannot be more energy reaching the surface of the member than the energy travelling through the smallest box surrounding the section (Wickström 2001). This can be seen from Figure 4.2 that shows the difference between the surface perimeter (full line) and the box perimeter (doted line) for an I-section and for an angle section.

Strictly speaking, the correction taking the shadow effect into account should apply only to the radiative part of the heat flux whereas, because it directly multiplies the totality of the heat flux in Equation 4.1, it applies also to the convective part of the flux.

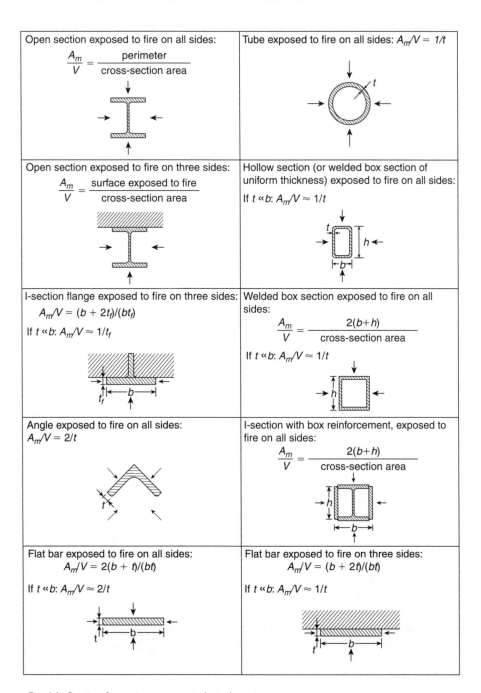

Open section exposed to fire on all sides: $$\frac{A_m}{V} = \frac{\text{perimeter}}{\text{cross-section area}}$$	Tube exposed to fire on all sides: $A_m/V = 1/t$
Open section exposed to fire on three sides: $$\frac{A_m}{V} = \frac{\text{surface exposed to fire}}{\text{cross-section area}}$$	Hollow section (or welded box section of uniform thickness) exposed to fire on all sides: If $t \ll b$: $A_m/V \approx 1/t$
I-section flange exposed to fire on three sides: $A_m/V = (b + 2t_f)/(bt_f)$ If $t \ll b$: $A_m/V \approx 1/t_f$	Welded box section exposed to fire on all sides: $$\frac{A_m}{V} = \frac{2(b+h)}{\text{cross-section area}}$$ If $t \ll b$: $A_m/V \approx 1/t$
Angle exposed to fire on all sides: $A_m/V = 2/t$	I-section with box reinforcement, exposed to fire on all sides: $$\frac{A_m}{V} = \frac{2(b+h)}{\text{cross-section area}}$$
Flat bar exposed to fire on all sides: $A_m/V = 2(b + t)/(bt)$ If $t \ll b$: $A_m/V \approx 2/t$	Flat bar exposed to fire on three sides: $A_m/V = (b + 2t)/(bt)$ If $t \ll b$: $A_m/V \approx 1/t$

Fig. 4.1 Section factor in unprotected steel sections

Table 4.1 Section factor and thermal massivity values for different sections

Term	Section factor (Massivity factor)	Thermal massivity
Equation	A_m/V	V/A_m
Unit	m^{-1}	m
Value for an open section	$2/t$	$t/2$
Value for a tube	$1/t$	t

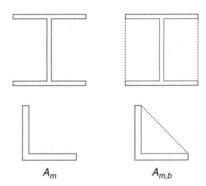

A_m $A_{m,b}$

Fig. 4.2 Surface area and box area

This approximation is justified by the fact that, for temperatures normally encountered in a fire, radiation is the dominant heat transfer mode to the section. Consequently, k_{sh} equals unity for cross sections with a convex shape, e.g. rectangular and circular hollow sections, in which the box area equals the surface area.

Generally speaking, k_{sh} is given by Equation 4.3.

$$k_{sh} = \frac{[A_m/V]_b}{[A_m/V]} = \frac{A_{m,b}}{A_m} \tag{4.3}$$

For the specific case of I-sections under nominal fire action, for example the ISO 834 or the hydrocarbon temperature-time curves, k_{sh} is given by Equation 4.4.

$$k_{sh} = 0.9\frac{[A_m/V]_b}{[A_m/V]} = 0.9\frac{A_{m,b}}{A_m} \tag{4.4}$$

Practically speaking, it is easy to use Equation 4.5 instead of Equation 4.1

$$\Delta\theta_{s,t} = \frac{A_m^*/V}{c_a\rho_a}\dot{h}_{net}\,\Delta t \tag{4.5}$$

where A_m^*/V is either based on A_m, $A_{m,b}$ or $0.9\,A_{m,b}$ depending on the situation. This simplifies, for example, the utilisation of graphical design aids.

As such, Equation 4.5 (or Eq. 4.1) does not directly yield the steel temperature at a specific time; it has to be integrated over time. A simple algorithm such as the one given below can be used to compute steel temperatures. This algorithm is explicit, which means that the temperature increase during a time step is calculated as a function of the values of all variables at the beginning of the time step. In order to ensure stability of the integration process, such an algorithm must be used with rather small time steps, not bigger than 5 seconds according to EN 1993-1-2. Other more refined algorithm can be written based on an implicit integration that allows using bigger time steps but, with modern computers, it is not a problem to use short time steps and the time required and the precision obtained with the explicit algorithm are normally satisfactory.

```
Data: Rho=7850; time=0; dtime=2; Tsteel=20; TimePrint=60
Data: h=25; eps=0.7
Read AmV_effective, FinalTime
Print AmV_effective
Do while (time < FinalTime)
        Call Sub_Csteel(Tsteel,Csteel)
        Call Sub_FireTemp(time,Tfire)
        Call Sub_Flux(h,eps,Tsteel,Tfire,hnet)
        Tsteel = Tsteel + (AmV_effective*hnet*dtime) / (Csteel*Rho)
        time = time + dtime
        If ( modulo(TimePrint,time) = 0 ) Print time, Tfire, Tsteel
EndDo
```

where:

Sub_Csteel	is a subroutine or a function that returns the value of the specific heat of steel as a function of the steel temperature, see Equation I.2 in Annex I.
Sub_FireTemp	is a subroutine that returns the value of the gas temperature as a function of time, see Equation 3.1, 3.2 and 3.3 for nominal fires.
Sub_Flux	is a subroutine that returns the value of the design value of the net heat flux as a function of the coefficient of convection, the emissivity, the gas temperature and the steel temperature, see Equation 3.14.

For a defined fire, it is quite convenient to do the integration of Equation 4.5 once for various values of the effective section factor A_m^*/V and to build design aids in the form of tables or graphs. For example, Table I.1 and Figure I.3 and I.4 presented in Annex I have been built for the ISO 834 standard fire. The explicit integration scheme has been used with a time step of 1 second.

The S shape of the curves that can be observed on Figure I.4 for temperatures around 735°C results from the peak in the specific heat of steel for this temperature, see Equation I.2. This Figure shows that, except for very massive sections, the steel temperature is higher than 700°C after 30 minutes. The temperatures obtained after 60 minutes are so high that it is impossible for an unprotected steel structure to have a fire resistance of one hour under standard fire exposure.

Figure I.4 shows the evolution of the temperature obtained after a given time as a function of the massivity factor. It has been suggested in some textbooks that a good mean for obtaining higher resistance times would be to select steel sections with higher massivity factors because the temperature increase is slower in sections that are more massive. This Figure shows that, for fire resistance times of 20 minutes and more, the steel temperature is hardly decreased if the massivity factor cannot be reduced below $200 \, \text{m}^{-1}$. A significant reduction of the temperature requires a reduction of the massivity factor to values lower than $100 \, \text{m}^{-1}$. Experience shows that, in reality, it is always more efficient to select a section that has higher mechanical properties, a higher plastic modulus for example in a section under bending, than to increase the thermal massivity.

4.2.2 Examples

4.2.2.1 Rectangular hollow core section

What is the effective section factor of a $200 \times 300 \times 10$ rectangular hollow core section exposed to the fire on four sides?

In this convex section, the box perimeter is equal to the surface perimeter and the section factor is based on the surface perimeter.

$$A_{m,b} = 2(200 + 300) = 1000 \text{ mm} = A_m$$

$$V = 200 * 300 - 180 * 280 = 9600 \text{ mm}^2$$

$$A_m^*/V = 0.104 \text{ mm}^{-1} = 104 \text{ m}^{-1}$$

Note: $V/A_m^* = 9.62 \text{ mm} \approx t$

4.2.2.2 I-section exposed to fire on 4 sides and subjected to a nominal fire

What is the effective section factor of a HE 200 A section heated on four sides and subjected to a nominal fire?

The effective section factor of this concave section has to be based on the box value of the perimeter and, because it is subjected to the nominal fire, the factor 0.9 has to be taken into account.

$$A_{m,b} = 2(h + b) = 2(0.190 + 0.200) = 0.780 \text{ m}$$

$$V = 53.8 \times 10^{-4} \text{ m}^2 \text{ (from catalogues)}$$

$$A_m^*/V = 0.9 \times 0.780/53.8 \times 10^{-4} = 130 \text{ m}^{-1}$$

Note: $V/A_m^* = 7.66 \text{ mm}$

4.2.2.3 I-section exposed to fire on 3 sides

What is the effective section factor of an IPE300 section exposed on three sides with a concrete slab on the top of the beam (any fire)?

The effective section factor of this concave section has to be based on the box value of the perimeter of the exposed part of the section.

$$A_{m,b} = 2h + b = 2 \times 0.300 + 0.150 = 0.750 \text{ m}$$

$$V = 53.8 \times 10^{-4} \text{ m}^2(\text{from catalogues})$$

$$A_m^*/V = 0.750/53.8 \times 10^{-4} = 139 \text{ m}^{-1}$$

Note: $V/A_m^* = 7.17 \text{ mm}$

The top surface of the steel section has not been taken into account for the evaluation of the boxed perimeter. This is to represent the fact that this surface is not in contact with the fire and that no heat transfer with the hot gas exists there. In fact, there will be a heat transfer on the top surface but in the direction from the steel section to the concrete slab and this transfer delays somewhat the temperature increase in the section. It will be explained in Chapter 5 how the fact that this heat sink effect is not taken into account in the thermal analysis is compensated for by a correction factor in the mechanical analysis.

4.3 Internal steelwork insulated by fire protection material

4.3.1 Principles

If the temperature distribution in the cross section is supposed to be uniform, the temperature increase during a time increment is given by Equation 4.6.

$$\Delta\theta_{a,t} = \frac{\lambda_p A_p/V}{d_p c_a \rho_a} \frac{(\theta_{g,t} - \theta_{a,t})}{(1 + \phi/3)} \Delta t - (e^{\phi/10} - 1)\Delta\theta_{g,t} \qquad (4.6)$$

$$\text{with } \phi = \frac{c_p \rho_p}{c_a \rho_a} d_p A_p/V$$

where:

λ_p is the thermal conductivity of the fire protection material,
A_p/V is the section factor for steel members insulated by fire protection material,
A_p is the appropriate area of fire protection material per unit length
        of the member,
V is the volume of the member per unit length,
$\theta_{g,t}$ is the ambient gas temperature at time t,
$\theta_{a,t}$ is the steel temperature at time t,
d_p is the thickness of the fire protection material,
c_a is the temperature dependant specific heat of steel,
ρ_a is the unit mass of steel,
Δt is the time interval,
$\Delta\theta_{g,t}$ is the increase of ambient gas temperature during the time interval Δt,
c_p is the temperature independent specific heat of the fire protection material,
ρ_p is the unit mass of the fire protection material,

The above equation is derived from the formulation of Wickström (1985) where the governing partial differential equation of the heat transfer inside the insulation layer

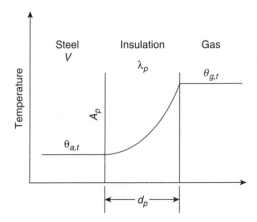

Fig. 4.3 Temperature in protected steelwork – schematic representation

was solved. Some simplifications of the solution of this 1D equation lead to the exponential correction factor. Strictly speaking, the approximation of the exact solution is valid for small values of the factor ϕ. This factor should normally not be higher than 1.5 but this limitation has not been specified in the Eurocode. A comprehensive discussion on various simple equations derived for evaluating the temperature increase in thermally protected steel member may be found in Wang (2004).

The design value of the net heat flux, and hence the coefficients for boundary conditions, do not appear in Equation 4.6 because the hypothesis behind this equation is that the surface temperature of the thermal insulation is equal to the gas temperature. The thermal resistance between the gas and the surface of the insulation is neglected. It is assumed that the temperature increase in the section is driven by the difference in temperature between the surface of the insulation, i.e. the gas temperature, and the steel profile, with only the thickness of the insulation providing a thermal resistance to conduction, see Figure 4.3.

Figure 4.4 shows how the section factor for steel members insulated by fire protection material is calculated under different configurations.

As Equation 4.1 or 4.5, Equation 4.6 has to be integrated over time in order to obtain the evolution of the temperature in the steel section as a function of time. EN 1993-1-2 recommends that the value of the time step Δt should not be taken as more than 30 seconds, a value deemed to ensure convergence even with an explicit method. Higher values could probably be taken into account with implicit methods, but the benefit in term of CPU time on modern computers would be marginal anyway; it is thus better to adhere to time steps not exceeding 30 seconds.

Figure 4.4, taken from Eurocode 3, shows that the section factor for sections insulated by a hollow encasement are based on the dimensions of the section, h and b, even if the encasement does not touch the section and, in that case, the surface that radiates energy to the steel section is the inside surface of the encasement. This approximation has been made in order to avoid the introduction of the distance between the section and the encasement as a new parameter in the design process.

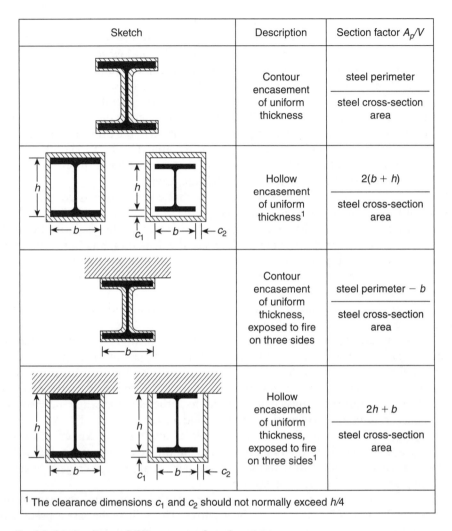

Sketch	Description	Section factor A_p/V
	Contour encasement of uniform thickness	$\dfrac{\text{steel perimeter}}{\text{steel cross-section area}}$
	Hollow encasement of uniform thickness[1]	$\dfrac{2(b + h)}{\text{steel cross-section area}}$
	Contour encasement of uniform thickness, exposed to fire on three sides	$\dfrac{\text{steel perimeter} - b}{\text{steel cross-section area}}$
	Hollow encasement of uniform thickness, exposed to fire on three sides[1]	$\dfrac{2h + b}{\text{steel cross-section area}}$

[1] The clearance dimensions c_1 and c_2 should not normally exceed $h/4$

Fig. 4.4 Section factor, A_p/V, in protected steel sections

This would complicate the design process especially when using design aids such as graphs and tables.

The thermal properties of the insulating material that appear in Equation 4.6 must have been determined experimentally according to ENV 13381-4 (2002). According to this standard, several short unloaded specimens as well as a limited number of loaded specimens, with various massivity factors and various protection thicknesses, are submitted to the standard fire. The thermal conductivity of the insulating material is back calculated from the recorded steel temperatures using the invert of Equation 4.6. The unit mass and the constant specific heat must be provided by the manufacturer of the product (if the specific heat is unknown, a value of 1000 J/kgK is assumed).

It is important to note that the thermal properties of the insulation determined according to ENV 13381-4 are directly applicable only to "I" or "H" type sections. Some corrections may be required if the product is to be applied on other section types such as "U" or "T" sections or rectangular and circular hollow sections. For reactive protection materials such as intumescent paint for example, additional tests may even be required if the product has to be applied on hollow sections.

If the thermal conductivity is considered as temperature dependent in the analysis of the results, a horizontal plateau at 100°C can be introduced in the time integration of Equation 4.6 when calculating the temperature evolution in the protected steel section, in order to take the evaporation of moisture into account. The duration of this plateau is a function of the thickness of the insulation. The evaporation of moisture can also be conservatively neglected in order to simplify the process.

There is also a possibility to consider the thermal conductivity as constant in the analysis of the results. In that case, the effect of the evaporation of moisture is implicitly taken into account in the effective thermal conductivity that is derived (but this conductivity is now a function of the thickness of the insulation and of the maximum steel temperature). A big error that must be avoided is to consider the values of the thermal properties derived at ambient temperature, typically for applications such as thermal insulation in buildings. This would lead to unsafe results in the fire situation because the thermal conductivity, for example, has a tendency to increase with increasing temperature in most insulating materials.

Generally speaking, and especially for reactive protections such as intumescent paints, the thermal conductivity is a function of the thickness of the protection.

What is very important is that the hypothesis made for the thermal conductivity of the insulating material when integrating Equation 4.6 to calculate the evolution of the steel temperature is consistent with the hypothesis made when analysing the experimental results for deriving this thermal conductivity.

The following algorithm shows a very simple example of an explicit integration scheme. It is consistent with the hypothesis made in the determination of the thermal conductivity that the temperature of the protection is equal to the average between the steel temperature and the gas temperature.

During the early stage of the fire, it may occur that the temperature increase in steel calculated by Equation 4.6 turns out to be negative. This will be the case especially for protection materials that have a high thermal capacity. In that case, the temperature increase in steel has to be set to 0 and the integration process continued in the next time steps. If the gas temperature is decreasing, a negative variation of the steel temperature can of course be accepted.

```
Data: Rhoa=7850; time=0; dtime=10; Ta=20; TimePrint=60
Data Tfire_previous = 20
Read ApV, dp, cp, Rhop, FinalTime
Print ApV, dp, cp, Rhop
Do while (time < FinalTime)
      Call Sub_Csteel(Ta,Ca)
      Phi = cp * Rhop *dp * Apv / (Ca * Rhoa)
      Call Sub_FireTemp(time,Tfire)
      dTfire = Tfire-Tfire_previous
```

```
      Tp = (Ta+Tfire)/2
      Call Sub_Lambdap(Tp,Lp,dp)
      dTa = Lp * ApV * dtime * (Tfire-Ta) / ( dp * Ca * Rhoa * (1+Phi/3))
      dTa = dTa - (exp(Phi/10) - 1 ) * dTfire
      if (dTfire>0).and.(dTa<0) then dTa = 0
      Ta = Ta + dTa
      time = time + dtime
      Tfire_previous = Tfire
      If ( modulo(TimePrint,time) = 0 ) Print time, Tfire, Ta
EndDo
```

where:

Sub_Csteel   is a procedure that returns the value of the specific heat of steel as
              a function of the steel temperature, see Equation I.2 in Annex I,
Sub_FireTemp is a procedure that returns the value of the gas temperature as a
              function of time, see Equation 3.1, 3.2 and 3.3 for nominal fires,
Sub_Lambdap   is a procedure that returns the value of the thermal conductivity
              of the thermal insulation as a function of the temperature and
              thickness of the protection.

It can be noticed that if the specific heat of the protection material c_p is neglected, the parameter ϕ that appears in Equation 4.6 is equal to 0 and Equation 4.6 reduces to Equation 4.7.

$$\Delta\theta_{a,t} = \frac{\lambda_p A_p}{d_p V} \frac{(\theta_{g,t} - \theta_{a,t})}{c_a \rho_a} \Delta t \qquad (4.7)$$

All parameters appearing in Equation 4.7 that define the steel section and the protection are grouped into one single factor according to Equation 4.8.

$$k_p = \frac{\lambda_p}{d_p} \frac{A_p}{V} \qquad (4.8)$$

For a defined fire, it is quite convenient to do the integration of Equation 4.7 once for various values of this factor k_p and to build design aids in the form of tables or graphs. For example, Table I.2 and Figure I.5 presented in Annex I have been developed for the ISO 834 fire (explicit integration, time step of 1 second). It has to be understood that this table and this Figure provide quite conservative results because the specific heat of the protection as well as any moisture that it may contain have been neglected. Table I.2 and Figure I.5 are based on the assumption that the factor k_p defined by Equation 4.8 does not depend on the temperature.

4.3.2 Examples

4.3.2.1 H section heated on four sides

What is the temperature of a HE 360 A protected by 20 mm of a protection material that has a constant thermal conductivity equal to 0.20 W/mK and is exposed to the standard fire of 1 hour duration? Compare the temperatures obtained in the section for a sprayed material with that of a boxed contour section.

Sprayed material

$A_p = 1.834$ m, taken from a producer's catalogue (if this quantity is not given in the catalogue, a good approximation would be given by $2h + 4b = 2 \times 0.35 + 4 \times 0.30 = 1.90$ m).

$V = 142.8$ m$^2 = 0.0143$ m^2

$k_p = (0.20/0.020)(1.834/0.0143) = 10$ W/m^{2}K $\times 128$ m$^{-1} = 1280$ W/m^{3}K

Interpolation in table I.2 between 520°C for $k_p = 1200$ and 650°C for $k_p = 2000$ yields a temperature of 533°C

Boxed contour

$A_p = 2h + 2b = 2 \times 0.35 + 2 \times 0.30 = 1.30$ m

$k_p = (0.20/0.020)(1.30/0.0143) = 10$ W/m^2 K $\times 90.9$ m$^{-1} = 909$ W/m^{3}K

Interpolation in table I.2 between 414°C for $k_p = 800$ and 520°C for $k_p = 1200$ yields a temperature of 443°C

4.3.2.2 H section heated on three sides

A mechanical analysis has shown that a steel member made of a HE 360 A will lose the load bearing capacity when its temperature reaches a value of 740°C.

What is the fire resistance if the section, heated on three sides by the standard fire, is left unprotected? What sort of thermal protection could yield a fire resistance of 2 hours?

Unprotected section

$A_m^*/V = 0.9 \ (2h + b)/V = 0.9 \times 1.00 \times 142.8 \times 10^{-4} = 130$ m^{-1}

The point (130 m^{-1}; 720°C) in Figure I.4 yields a fire resistance time of approximately 22 minutes. A double linear interpolation in Table I.1 yields 21.7 minutes.

Protected section

Interpolation in Table I.2 between 654°C for $k_p = 800$ and 734°C for $k_p = 1200$ indicates that k_p should have a value of 1130 for limiting the temperature to 720°C during 120 minutes.

If a sprayed protection is chosen, then $A_p = 1.834 - 0.30 = 1.534$ m and the insulating material should be characterised by $\lambda_p/d_p = 1130 \times 0.0143/1.534 = 10.5$ W/m^{2}K. A protection material with a thermal conductivity of 0.15 W/mK, for example, should have a thickness of $0.15/10.5 = 0.0142$ m ≈ 15 mm to yield 2 hours fire resistance.

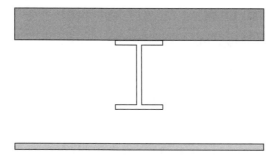

Fig. 4.5 Heat screen under a steel beam

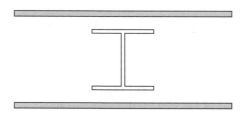

Fig. 4.6 Steel column enclosed between vertical heat screens

4.4 Internal steelwork in a void protected by heat screens

Eurocode has special rules to address the scenario when voids (gaps) are present between the structural members and protection membrane (heat screen). One situation is a steel beam that has a floor on its top and has fire protection underneath through a heat screen, see Figure 4.5. The screen must not be in direct contact with the member. Another situation is a steel column in a void that has vertical heat screen on both sides, also without a direct contact between the member and the screens, see Figure 4.6. In each case, the properties and performance of the heat screens should have been determined using a test procedure conforming with ENV 13381-1 or ENV 13381-2 as appropriate.

Eurocode 3 allows calculating the temperature development in the steel member taking the ambient gas temperature $\theta_{g,t}$ as equal to the gas temperature measured in the void during the test. The temperature evolution in the steel member is determined according to one of the equations described previously in Section 4.1, if the steel member is unprotected, or Section 4.2 if there is a thermal insulation around the steel member. Of course, the evolution of the temperature in the void does not follow the evolution of a nominal fire and it is impossible to build design aids for heat screens in general. Design aids can be built only if they are in each case relevant to one particular heat screen.

If a particular heat screen has been tested experimentally and has proved to comply with all three criteria R, E and I during a certain amount of time, no verification of the steel member is needed because the temperature increase in the void cannot be higher than 140°C, which is the criterion for thermal insulation. If, on the contrary, the false ceiling has only satisfied to the stability criterion R, the steel member must be verified because the temperature increase in the void can be significant.

4.5 External steelwork

4.5.1 General principles

This section refers to steel members, either columns or beams, which are located outside the envelope of the building where the fire is taking place. These members can nevertheless be influenced and heated by the fire in different manners:

- By the radiative heat flux from the openings of the fire compartment, e.g. the windows.
- By the flames projecting from the openings of the compartment.

A member that is not engulfed in flames is heated by radiation from all openings in that side of the fire compartment and by radiation from all the flames that project from all these openings.

A member that is engulfed in flames is heated by convection and radiation from the engulfing flame as well as by radiation from the opening from which the engulfing flame projects. The effect of the other flames and openings is neglected.

The steel member, in its turn, losses energy by radiation and convection either to the ambient environment if not engulfed in flame or to the engulfing flame when relevant.

If required, it is possible to protect the external steel member from the radiative effect of the fire by heat screens that are non-combustible and have a fire resistance of at least EI 30 according to EN ISO 1350-2. It is assumed that no heat transfer takes place to those sides that are protected by heat screens.

The temperature of the steel member is determined from an equation expressing the steady state heat balance between the energy received by the member from the flames and openings and the energy leaving the member. Practical application of the method requires considering information and equations that are presented partly in Annex B of Eurocode 1 – Part 2, for determining the maximum temperature in the compartment, the size and temperature of flames from the openings and radiative and convection parameters, and partly in Annex B of Eurocode 3 – Part 2, for the heat balance equations. Amazingly enough, the Annex B of Eurocode 1 is informative whereas the Annex B of Eurocode 3 is normative.

These equations are numerous and complex in some cases. They require no particular comment. The example presented in the following section may help the reader finding his way more easily through these equations than by reading simply the text of the Eurocodes.

4.5.2 Example

A fire compartment is rectangular in plane with dimensions 6 by 12 m^2 and 3 m high. There is one single window of dimension $w_t \times h = 3 \times 1.5$ m^2 in the shortest wall. The design fire load $q_{f,d}$ is 500 MJ/m^2.

A steel column is at a distance of 0.50 m from the façade, right in the centre of the window. It is made of a 150×150 mm square hollow section.

Determine the resulting temperature in the steel column under steady state situation.

From Annex B of Eurocode 1
Because there is a window only in one wall and there is no core in the fire compartment, the ratio D/W is given by:

$D/W = W_2/w_t = $ Width of the wall perpendicular to the wall of the window/window
width $= 12/3 = 4$.
Area of the window: $A_v = 1.5 \times 3 = 4.5\,\mathrm{m}^2$
Total area of enclosure: $A_t = 2(6 \times 12 + (6 + 12)3) = 252\,\mathrm{m}^2$
The opening factor of the fire compartment O is equal to $A_v\sqrt{h}/A_t = 0.022\,\mathrm{m}^{0.5}$

Because there is no window on opposite sides of the compartment, the calculation is done with no forced draught ventilation.
The rate of heat release is given by:

$$Q = \min\left(\frac{A_f q_{f,d}}{\tau_F}; 3.15\left(1 - e^{-0.036/o}\right) A_v\sqrt{\frac{h}{D/W}}\right)$$

with A_f floor area $= 6 \times 12 = 72\,\mathrm{m}^2$
   τ_F free burning fire duration, assumed to be 1 200 seconds
The first term in the parenthesis is the rate of heat release in a fuel-controlled situation. It is equal to 30 MW.
The second term in the parenthesis is the rate of heat release in an air-controlled situation. It is equal to 7 MW.
The rate of heat release is thus equal to 7 MW.
The temperature of the fire compartment is given by:

$$T_f = T_0 + 6000\left(1 - e^{-0.1/o}\right)O^{0.5}\left(1 - e^{-0.00286\Omega}\right) = 857°C$$

with $\Omega = \dfrac{A_f q_{f,d}}{\sqrt{A_v A_t}} = \dfrac{72 \times 500}{\sqrt{4.5 \times 252}} = 1069\,\mathrm{MJ/m}^2$

The flame height, if the wind velocity is equal to 6 m/s, is given by:

$$L_L = 1.9\left(\frac{Q}{w_t}\right)^{2/3} - h = 1.85\,\mathrm{m}$$

Flame width $=$ window width $= 3$ m.
Flame depth $= 2/3$ window height $= 1\,\mathrm{m} < 1.25\,w_t$, see Figure 4.7.
Horizontal projection of flames, $L_h = h/3 = 0.5\,\mathrm{m}$ (there is a wall above the window).
Flame length along axis, $L_f = L_L + h/2 = 1.85 + 0.75 = 2.60\,\mathrm{m}$.

The flame temperature at the window is given by:

$$T_w = T_0 + \frac{520}{1 - 0.4725 \times 1.0} = 1006°\,C\ (1279\,K)$$

because $L_f w_t/Q = 1.11 > 1$ and is replaced by the value of 1.
The emissivity of the flame at the window (ε_f) is taken as 1.0

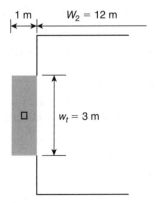

Fig. 4.7 Plan view of the region near the window

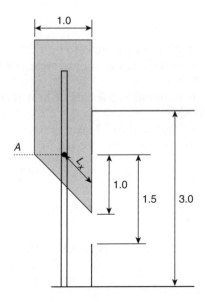

Fig. 4.8 Elevation of the region near the window

Axis length from the window to the point where the temperature analysis is made, see point A on Figure 4.8: $L_x = 0.707$ m.

The flame temperature at point A is given by:

$$T_z = T_0 + (T_w - T_0)(1 - 0.4725\, L_x w_t / Q) = 865°C$$

The emissivity of flames is taken as $1 - e^{-0.3\, d_f} = 0.259$

where d_f is the flame thickness $= 1$ m

The convective heat transfer coefficient is given by:

$$\alpha_c = 4.67(1/d_{eq})^{0.4}(Q/A_v)^{0.6} = 13 \, \text{W/m}^2\text{K}$$

From Annex B of Eurocode 3
As the column is engulfed in flames, its average temperature T_m is determined as the solution of the following heat balance equation:

$$\sigma T_m^4 + \alpha T_m = I_z + I_f + \alpha T_z \qquad (4.9)$$

where:

σ is the Stefan-Boltzmann constant, $5.67 \; 10^{-8} \, \text{W/m}^2\text{K}^4$,
α is the convective heat transfer coefficient, $13 \, \text{W/m}^2\text{K}$,
T_z is the flame temperature, $865°\text{C}$ ($1138 \, \text{K}$),
I_z is the radiative heat flux from the flame,
I_f is the radiative heat flux from the window.

Radiation from the flame.

$$I_z = \frac{(I_{z,1} + I_{z,2})d_1 + (I_{z,3} + I_{z,4})d_2}{2(d_1 + d_2)}$$

d_1, width of the member perpendicular to the window $= 0.15 \, \text{m}$.
d_2, width of the member parallel to the window $= 0.15 \, \text{m}$.
$\varepsilon_{z,1}$, emissivity of the flame on side 1 of the member, i.e. perpendicular to the window $= 1 - e^{-0.3 \times 1.425} = 0.348$

    => $I_{z,1} = 0.348 \times 5.67 \times 10^{-8} \times 1138^4 = 33.09 \, \text{kW/m}^2$
    => $I_{z,2} = I_{z,1} = 33.09 \, \text{kW/m}^2$

$\varepsilon_{z,4}$, emissivity of the flame on side 4 of the member, i.e. opposite to the window $= 1 - e^{-0.3 \times 0.35} = 0.100$

    => $I_{z,4} = 0.100 \times 5.67 \times 10^{-8} \times 1138^4 = 9.48 \, \text{kW/m}^2$
    $\varepsilon_{z,3} = \varepsilon_{z,4}$
    => $I_{z,3} = 0.100 \times 5.67 \; 10^{-8} \times 1279^4 = 15.17 \, \text{kW/m}^2$
$I_z = (2 \times 33.09 + 9.48 + 15.17)/4 = 22.71 \, \text{kW/m}^2$

Radiation from the opening

$$I_f = \phi_f \varepsilon_f (1 - \alpha_z) \sigma T_f^4$$

$$\phi_f = \frac{(\varphi_{f,1} + \varphi_{f,2})d_1 + \varphi_{f,3}d_2}{2(d_1 + d_2)}$$

$\varphi_{f,3}$, view factor from the window to side 3 of the member, i.e. the side of the member facing the window, $= 2 \times 0.239 = 0.478$ according to equation G.2 of Eurocode 1, with the window divided in 2 zones of $1.5 \times 1.5 \, \text{m}^2$ at a distance of $0.35 \, \text{m}$ from the member.

$\varphi_{f,2}$, view factor from the window to side 2 of the member, i.e. one side perpendicular to the plane of the window, $= 0.158$ according to equation G.3 of Eurocode 1, with $w = 1.425$ m, $s = 0.5$ m and $h = 1.5$ m.

$\varphi_{f,1} = \varphi_{f,2}$, owing to symmetry.

$$\phi_f = \frac{2 \times 0.158 + 0.478}{4} = 0.199$$

$\varepsilon_f = 1.0$
$\alpha_z = (\varepsilon_{z,1} + \varepsilon_{z,2} + \varepsilon_{z,3})/3 = 0.265$
$T_f = 857°C \ (1130\,K)$
$I_f = 0.199 \ (1 - 0.265) \ 5.67 \ 10^{-8} \times 1130^4 = 13.55 \ kW/m^2$

Incident heat flux
$22\ 710 + 13\ 550 + 13 \times 1138 = 51\ 054 \ W/m^2$

This flux is the quantity that appears on the right hand side of Equation 4.9. It is straightforward to solve this equation. This yields:

$T_m = 912\,K \ (639°C)$

This temperature is the temperature that will be established in the member in the steady state situation. No indication is given by this method concerning the time that will elapse before this situation exists. Of course, it would be easy to write the transient equivalent of Equation 4.9 and to integrate this resulting equation over time, which would yield the evolution of the temperature in the steel member as a function of time.

Chapter 5

Mechanical Analysis

5.1 Level of analysis

5.1.1 *Principles*

The response of a structure exposed to fire can be analysed at different levels. It is the responsibility of the designer to select the level of analysis. The three possibilities are:

Global structural analysis

If the structure is rather simple or, in case of a complex structure, if a sufficiently sophisticated tool is available for the analysis, it is possible to consider the entire structure as a whole and to analyse it as a single object.

Member analysis

On the opposite, the structure can be seen as the assembly of members, here defined as load bearing elements limited in their dimensions either by a support with the foundation or by a joint with other elements. Typically, the word "member" designates a beam, a column, a floor, etc. It is possible to analyse the structure as a collection of a number of individual elements, the fire resistance of the structure being taken as the shortest fire resistance of all the members.

Sub structure analysis

This is the intermediate solution between the aforementioned limit cases; any part of the structure that is bigger than an element is a substructure.

It has to be noticed that the same choice is in fact also made for the design at room temperature:

- A structure can be entirely represented (discretised) as a single object and the effects of actions determined in this object, usually by a computer analysis program.
- Yet, very large structures, such as the Eiffel Tower in Paris, have been designed as an assembly of elements, the resistance of each of them being verified individually.
- In an industrial hall made of parallel one storey one bay portal frames with purlins spanning from frame to frame, a usual procedure would be:
  - design the purlins as individual elements,
  - design the frames as separate substructures, i.e. one representative frame is considered individually (no 3D interaction with the other frames) but analysed

as a whole and not as the addition of two columns and one beam, which means, for instance, that moment redistribution is considered within the frame,
–   design the bracing system also as a substructure, for example as a statically determinate truss girder.

The problem is somewhat more complex in case of fire because of indirect fire actions, i.e. these variations of axial forces, shear forces and bending moments resulting from restraint to the thermal movements.

In a global structural analysis, all indirect fire actions developing in the structure during the course of the fire must be considered.

In a substructure analysis, the conditions in terms of supports and/or actions acting at the boundary of the substructure are evaluated at time $t = 0$, i.e. at the beginning of the fire, and are considered to remain constant during the entire duration of the fire. Indirect fire actions can nevertheless develop within the substructure.

In a member analysis, boundary conditions are also fixed to the value that they have at the beginning of the fire, but no indirect fire action is taken into account in the member, except those resulting from thermal gradients. In fact, the only cases where the effects of thermal gradients have been recognised to have significant effects on the fire resistance of simple members are the cantilever or simply supported walls or columns submitted to the fire on one side only. In these cases, the important lateral deflections induced by the thermal gradient may generate significant additional bending moments due to second order effects. This can lead to premature failure, either by yielding of the material at the base of the column, by general buckling of the column, or even by loss of equilibrium of the foundation. This is clearly a case where the effects of thermal gradients have to be taken into account, even in a member analysis.

It should be noted that significant thermal expansion will in fact be present in the structure and this should be accounted for in the analysis through the discretisation of the structure into elements and/or substructures in such a way that these hypotheses on the constant boundary conditions are reasonable and correspond at least to a good approximation of the real situation. Designing, for example, as a simple element a beam of an underground car park that is very severely restricted against thermal elongation by the surrounding ground and neglecting the increase of axial compression force that will for sure arise in reality, would not represented a sound approximation of the actual boundary conditions present during fire exposure.

5.1.2 Boundary conditions in a substructure or an element analysis

No precise recommendation is given in the Eurocode concerning the way to define the boundary conditions at the separation between an element or a substructure and the rest of the structure.

The following procedure is recommended by the authors for selecting the boundary conditions in a substructure or an element. It is here explained for a substructure, but the same would hold for an element.

1.   The effects of action in the whole structure must be determined at time $t = 0$ under the load combination in case of fire that is under consideration.
2.   The limits of the substructure have to be chosen. The choice is made with the contradictory objective that the substructure becomes as simple as possible, but

at the same time the hypothesis of constant boundary conditions during the fire must represent a good approximation of the real situation, with respect to the thermal expansion that exists in reality. The choice of the limits of the substructure is of course highly dependent of the location of the fire. Engineering judgement is necessary.

3. All the supports of the structure that belong to the substructure have to be taken into account as supports of the substructure.

4. All the external mechanical loads that are applied on the substructure in case of fire have to be taken into account as acting on the substructure.

5. For each degree of freedom existing at the boundary between the substructure and the rest of the structure, an appropriate choice has to be made in order to represent the situation as properly as possible. The two possibilities are:
   (a) the displacement (or the rotation) with respect to this degree of freedom is fixed, or
   (b) the force (or the bending moment) deduced from the analysis of the total structure computed in step 1 is applied.

   These two possibilities are exclusive because it is not possible to impose simultaneously the displacement and the corresponding force at a degree of freedom. Whatever the choice, these restrictions on the displacements and these forces applied at the boundaries will remain constant during the fire.

6. A new structural analysis is performed at room temperature on the substructure or the element that has been defined and it yields the effects of actions that have to be taken into account in the substructure or in the element.

7. In a substructure analysis, the indirect fire actions that could develop within the substructure have to be taken into account, whereas this is not the case for an element analysis.

This procedure allows finding ones way through the most complex cases but, as the examples below will demonstrate, some of the steps are trivial or omitted in more simple cases. Two examples are given in chapter 8 that illustrate this procedure, first for a continuous beam, and then for a multi-storey framed structure.

5.1.3 Determining $E_{fi,d,0}$

It has been mentioned in Section 5.1.2 that the effects of actions at time $t = 0$, noted $E_{fi,d,0}$, have to be determined in order to perform a member or a substructure analysis. No indication is given in the Eurocodes concerning the analysis method that has to be used to determine these effects of action.

In practice, this is normally done by an elastic analysis because it is reasonable to assume that the structure will exhibit very little if any plasticity under the design loads in case of fire. Indeed, if the situation prevailing at the beginning of the fire is compared to the situation that has been taken into account for the design of the structure under normal conditions, the design values of the mechanical loads as well as the partial safety factor dividing the resistance of the material are lower. A steel structure that has been designed in order to sustain in normal conditions a design load equal to $1.35G + 1.50Q$ with a resistance of $f_y/\gamma_{M,1} = f_y/1.15$, for example, will exhibit very little plasticity at the beginning of the fire if the load is only $1.00G + 0.50Q$ and the full resistance $f_y/\gamma_{M,fi} = f_y$ can be mobilised.

Because the effects of actions are determined at time $t = 0$, the stiffness of the material at room temperature is of course taken into account.

If the structure is simple, the analysis is trivial but, if the structure is complex, it is possible to use one of the numerous numerical tools developed for the analysis of structures at ambient temperature.

5.2 Different calculation models

5.2.1 General principle

Three different calculation models can be applied for the determination of the fire resistance of a structure or an element. They differ very much in their complexity, but also in their field of application and in what they can offer. It is important, before a choice is made, that these differences are clearly identified.

These calculation models are discussed hereafter, from the simplest to the most complex one.

5.2.1.1 Tabulated data

Tabulated data directly give the fire resistance time as a function of a limited set of simple parameters, e.g. the concrete cover on the reinforcing bars in a reinforced concrete section or the thickness of insulation in a steel section, the load level, or the dimensions of the section. Such a model is thus normally easy to use.

Tabulated data are not based on equilibrium equations, but result from the empirical observation of either experimental test results or results of calculations made by more refined models. The tabulated data methods aim at representing these results with the best possible fit.

The name "tabulated" for this group of calculation models comes from the fact that the results are usually presented in the form of multi-entry tables. It has to be emphasized that some methods, even if they are presented in the form of analytical equations, belong in fact to the group of the tabulated data calculation models if they are not based on equilibrium conditions but, on the contrary, simply represent a best fit correlation with results obtained by another way.

The main limitations of tabulated data are:

- Tabulated data exist only for simple elements at present.
  Theoretically speaking, nothing speaks against establishing tabulated data for more complex structures, for example for single storey one bay frames, but the effort needed to establish these tables would be high, and the number of input parameters would probably increase to a point that much of the simplicity of the method would be lost.
- Such tabulated data have been established so far only for the standard fire curve, namely the ISO fire curve or its equivalent.
  It would in fact be totally impossible, even for the simplest elements, to build tabulated data encompassing all the possible natural fire curves that could exist, simply because the number of these curves is infinite. It has yet to be mentioned that some recent developments have been made in order to establish tabulated data

in case of particular parametric fire curves, namely those recommended in Annex A of Eurocode 1. In this particular case, it is possible to establish tabulated data in which the duration of the ISO fire that is usually present in the traditional tabulated data has been replaced by other factors, such as the fire load and the opening factor of the compartment for example. Such tables could, for example, allow verifying that a steel element with a defined thermal massivity, and a defined load level, can survive the parametric fire provided that the fire load does not exceed a certain value.

Whereas tabulated data are extensively used for concrete and composite steel-concrete structures, no tabulated data is presented in Eurocode 3, probably because the simple calculation model is of rather simple application. In the past, a nommo-gram had been published by the ECCS (1983) for unprotected elements and elements protected by a lightweight insulating material. It related graphically the fire resistance time to the standard fire, the thermal massivity of the section, the load level and, if relevant, the amount of thermal protection.

5.2.1.2 Simple calculation models

Simple calculation models must be simple enough to be applied in everyday practice without using sophisticated numerical software.

They must be based on equilibrium equations.

The ability of the element or the structure to sustain the applied loads is verified taking into account the elevation of the temperature in the material. Usually, the simple calculation models for steel elements are the direct extrapolation of models used for normal design at room temperature, in which the yield strength and Young's modulus of steel have been adapted in order to reflect the decrease induced by the increase of the temperature in steel. Some modifications may be introduced in the model in order to take into account certain aspects specific to the fire situation.

On the contrary to tabulated data, simple calculation models are applicable for any temperature-time fire curve, provided that the adequate material properties are known. It is for example essential to know whether any property determined during first heating is reversible during the cooling phase that will occur in a natural fire curve. Attention must also be paid to verify that the heating or cooling rate in the material belongs to the range for which the material properties have been determined.

Note: It has to be noted that Eurocode 3 gives no indication about the properties of steel during or after cooling. Information has to be taken from the literature, for example Kirby et al. 1986.

The main field of application of simple calculation models is the element analysis, although some simple substructures could theoretically also be analysed.

5.2.1.3 Advanced calculation models

Advanced calculation models are those sophisticated computer models that aim at representing the situation as close as possible to the scenario that exists in real structure.

Such models must be based on acknowledged and recognised principles of the structural mechanics.

Table 5.1 Relation between calculation models and division of the structure

	Element	Sub-structure	Structure
Tabulated data	++	−	− −
Simple calculation model	++	+	−
Advanced calculation model	+	++	++

It has to be emphasized that the fact to program a simple calculation method in a computer in order to facilitate its utilisation does not make it an advanced calculation model.

Advanced calculation models are applicable with any temperature-time fire curve provided that the appropriate material properties are known. They can be used for the analysis of entire structure because they take indirect fire actions into account.

More information about advanced calculation models is presented in Chapter 7.

5.2.2 Relations between the calculation model and the part of the structure that is analysed

A confusion is often made between the three calculation models, namely tabulated data, simple calculation models and advanced models on one hand (see Section 5.2.1), and the three levels of division of the structure, namely the element analysis, the sub-structure analysis and the structure analysis on the other hand (see Section 5.1.1). Although these are two different aspects of the question, there are of course some clear links between these two aspects. Table 5.1 which illustrates the relation between calculation models and division of structure may clarify the situation.

The table shows that:

• Tabulated data are to be used mainly with simple elements. Although one could imagine that such tabulated data be developed for simple substructures, this has actually not yet been done. Complete structures cannot be analysed by means of tabulated data.
• Simple calculation models can certainly be used for simple elements and, to some extent, for simple substructures. The analysis of complete structures can normally not be undertaken with simple calculation models.
• Advanced calculation models are the tool of choice for the analysis of complete structures or, if the time for the analysis has to be reduced, for substructures. They can also be used for simple elements but, in many cases, the simplicity, greater availability and user friendliness of simple calculation models, will lead to the decision to use these simple models when possible.

5.2.3 Calculation methods in North America

The most widely used approach for evaluating fire resistance in North America is based on manufacturer listings or through the use of prescriptive based simple calculation methods.

ASCE/SFPE 29 contain a number of simplified equations for determining fire resistance of steel structural members. These empirical equations are derived based on the results of standard fire resistance tests carried out on steel structural assemblies under standard fire exposure. These empirical methods often utilize factors such W/D ratios, where W is defined as the weight per unit length of the steel member (in lbs), and D is defined as the inside perimeter of the fire protection (in inches), for defining fire resistance. The methods are based on the presumption that the rate of temperature rise in a structural steel member depends on its weight and the surface area exposed to heat. For calculating the fire resistance of tubular column sections A/P factors are used, where A is the section area (in square inches) and P is the section perimeter (in inches). Specific values of W/D and A/P ratios for various steel sections and configurations (three or four side exposure types) are listed in Tables of AISC Manual (AISC 2005) and Appendix A of the AISC Steel Design Guide 19 (Ruddy et al. 2003).

The rational approach for evaluating fire resistance of steel structural members design is contained in ASCE manual of fire protection (Lie 1992) and SFPE Handbook of Fire Protection Engineering (SFPE, 2002). However, these sources have only limited information and do not contain details for the advanced calculation methods. The AISC manual of construction contains some discussion on rational fire design principles and refers to Eurocode 3(2003) for calculation methodologies. Thus, it is possible for designers in North America and other parts of the world to apply Eurocode methodologies for evaluating fire resistance and gain acceptance from regulatory officials. In such scenarios, the relevant high temperature properties for structural steel and insulation should be used. The high temperature material properties for structural steel and room temperature properties of insulation are listed Annex I and Annex II.

In evaluating fire resistance, often critical temperature in steel is used to define failure of a steel structural member. A critical temperature is defined as the temperature at which steel looses 50% of its room temperature strength (often yield strength). The rational for using critical temperature to define the failure is based on the premise that the loading on the structure, under fire conditions, is about 50% of its full capacity. In North America the critical temperatures limit commonly used are 538°C for steel columns and 593°C for steel beams (similar to temperature acceptance criteria adopted in ASTM E119) regardless of the loads applied to structural members. This is in contrast to European practice, see Section 5.7, where critical temperatures for steel members are specified depending on the so-called applied load level, or load ratio, i.e. critical temperatures are dependent on both the type of the structural member and the load level. The critical temperatures, however, are independent of time, and also, independent of the shape or size of the steel section.

5.3 Load, time or temperature domain

The stability analysis can be performed through different approaches mentioned in the Eurocode: namely in the time domain, in the load domain and, in some cases, in the temperature domain. These possibilities are illustrated on Figure 5.1 and Figure 5.2 for a simple case in which the applied load, in fact the effect of action $E_{fi,d}$, is constant during the fire and the element is characterised by a single temperature, $\theta_{structure}$.

Figure 5.1 refers to the case of a nominal fire in which the fire temperature, θ_{fire}, is continuously increasing. The temperatures in the structure, $\theta_{structure}$, will therefore also be a continuously increasing as a function of time and, although this will not

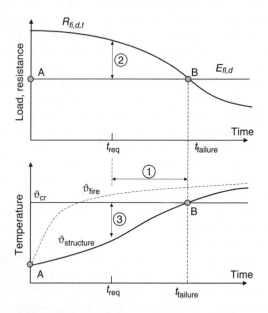

Fig. 5.1   Load, time or temperature domain for a nominal fire

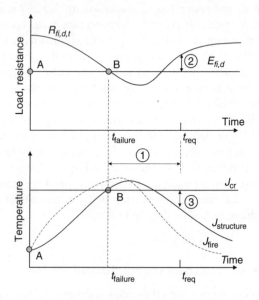

Fig. 5.2   Load, time or temperature domain for a natural fire

be demonstrated theoretically, it will be assumed that this induces a continuously decreasing load bearing capacity, $R_{fi,d,t}$.

The situation is different in the case of a natural fire in which the fire temperature has an increasing phase systematically followed by a cooling down phase, see Figure 5.2. The temperature in the structure will follow a similar evolution, although with a time delay. For steel structures, the load bearing capacity of the structure that could be calculated at different moments in time produces a pattern as shown on Figure 5.2, with a first phase where the load bearing capacity decreases as a function of time, and a second phase when the structure recovers its load bearing capacity, mainly because steel recovers its strength, either totally or partially, when cooling down to ambient temperature.

In each case, t_{req} noted on the Figures is the required fire resistance time of the structure.

The situation at the beginning of the fire is represented by point A on the Figures and, if the analysis is performed by the advanced calculation model, the method, i.e. the software normally, will usually track the evolution of the situation of the structure until point B when failure occurs (most computer software indeed perform a transient step by step analysis). This means that the curve showing the evolution of the load bearing capacity is not known to the designer.

If, on the contrary, the analysis is performed by the simple calculation model, there are different manners to verify the stability, normally referring to one of the points of this curve. The three verification possibilities are:

1. In the time domain.
    It has to be verified that the time of failure t_{failure} is higher than the required fire resistance time t_{req}. This is expressed by Equation 5.1 and corresponds to the verification 1, satisfied on Figure 5.1 but not satisfied on Figure 5.2.

    $$t_{\text{failure}} \geq t_{\text{req}} \tag{5.1}$$

2. In the load domain.
    At the required time in the fire t_{req}, it is verified that the resistance of the structure $R_{fi,d,t}$ is still higher than the effect of action $E_{fi,d}$. This is expressed by equation 5.2 and corresponds to the verification 2 on Figures 5.1 and Figure 5.2.

    $$R_{fi,d,t} \geq E_{fi,d} \quad \text{at } t = t_{\text{req}} \tag{5.2}$$

    This verification is proposed as the standard method in Eurocode 3.

    It can be shown that, in the case of a fire with no decreasing phase, the fact that Equation 5.2 is satisfied guarantees that Equation 5.1 is also satisfied, see Figure 5.1. On the other hand, in the case of a fire with a cooling down phase, it can happen at some stage that Equation 5.2 is satisfied whereas Equation 5.1 is not satisfied, see Figure 5.2.

3. In the temperature domain.
    At the required fire resistance time t_{req}, it has to be verified that the temperature of the structure $\vartheta_{\text{structure}}$ is still lower than the critical temperature ϑ_{cr}, i.e. the temperature that leads to failure. This is expressed by Equation 5.3 and corresponds to the verification 3 on Figures 5.1 and 5.2.

    $$\vartheta \leq \vartheta_{cr} \quad \text{at } t = t_{\text{req}} \tag{5.3}$$

    This verification is a particular case of the verification in the load domain, only possible when the stability of the structure is depending on a single temperature, which is the case in steel elements under uniform temperature distribution. It can also happen for natural fires that Equation 5.3 is satisfied whereas Equation 5.1 is not.

It appears thus from the above discussion that, in the case of a natural fire, a single verification made in the load or in the temperature domain is not sufficient if the required fire resistance time is higher than the time of minimum load bearing capacity, i.e., usually, the time of maximum temperature.

It is possible to make the verification in the load domain several times at different times of fire until, after an iterative process, the time is found where the resistance of the structure is equal to the applied load. This time is, by definition, the failure time and it can be compared to the required fire resistance time.

It is yet simpler and possible to make one single verification if $R_{d,fi}$ in Equation 5.2 or θ in Equation 5.3 are not taken systematically at $t = t_{req}$, but at the time of the maximum steel temperature.

The rest of this chapter is presented based on the verification in the load domain. This type of verification has indeed several advantages.

1. It is easy to use; because the verification is at a given time; the steel temperature and hence the material properties are known and can be used for the evaluation of the load bearing capacity.
2. It is applicable for any type of effect of actions whereas, as will be explained in Section 5.7, verification in the temperature domain is possible only in a limited number of cases.
3. It produces a safety factor that is similar to the one that engineers and designers have been using for years at room temperature, namely the ratio between the applied load and the failure load. On the other hand, verification in the temperature domain yields a safety factor in degrees centigrade that does not provide much in term of practical consequences. A verification in the time domain may even be more confusing because, with the tendency of standard fire curves to level off at nearly constant temperatures after a certain period of time, they can yield the false impression of a very high level of safety because the calculated time of failure is significantly longer than the required fire resistance time, simply because the temperature of the structure changes very slowly, whereas a small variation in the applied load or in the heating regime would decrease the fire resistance time very dramatically close to the required resistance time.

5.4 Mechanical properties of carbon steel

For ambient design of building members at the ultimate limit state, carbon steel is usually idealised either as a rigid-plastic material, for evaluating plastic bending capacity of sections for example, or as an elastic-perfectly plastic material, for instability problems such as buckling for example.

At elevated temperatures, the shape of the stress-strain diagram is modified. The model recommended by EN 1993-1-3 is an elastic-elliptic-perfectly plastic model, plus a linear descending branch introduced at large strains when this material is used in advanced calculation models. The first part of the stress-strain relationship is schematically represented by the continuous curve O-A-B on Figure 5.3.

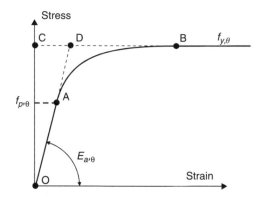

Fig. 5.3 Stress–strain relationship (schematic) for steel

The strain–strain relationship at elevated temperature is thus characterised by 3 parameters:

- The limit of proportionality $f_{p,\theta}$
- The effective yield strength $f_{y,\theta}$
- The Young's modulus $E_{a,\theta}$

Note: the strain for reaching the effective yield strength, point B on the Figure, is fixed at 2%.

The Eurocode contains a table that gives the evolution of these properties as normalized by the relevant property at room temperature, namely:

- $k_{p,\theta} = f_{p,\theta}/f_y$
- $k_{y,\theta} = f_{y,\theta}/f_y$
- $k_{E,\theta} = E_{a,\theta}/E$

This table is reproduced in Annex II of this book.

5.5 Classification of cross-sections

Stocky steel members are able to support a significant degree of rotation and/or compression without any local deformation and to develop the full plastic capacity of the section. Steel members made of thin plates, on the contrary, suffer from severe local deformations, possibly at load levels that are below the elastic capacity of the section.

In the philosophy of the Eurocodes, steel sections are sorted into 4 different classes with respect to the susceptibility to local buckling.

1. Class 1 sections are the stockiest sections. They are able to develop the full plastic capacity and this capacity is maintained for very large deformations. The ductility is sufficient to allow a redistribution of the bending moments along the length of the members by formation of plastic hinges during a loading to failure.

2.   Class 2 sections are also able to develop the full plastic capacity, but this capacity cannot be maintained for large deformations. Plastic redistribution along the members is not possible with such a section.
3.   Class 3 sections are able to develop the full elastic capacity but cannot reach the plastic capacity.
4.   Class 4 sections are the thinnest sections. In these sections, local buckling occurs for load levels that are below the full elastic capacity of the section.

At room temperature, the classification of a section depends on different parameters such as:

- The geometric properties of the section, via the slenderness of the plates that form the section. The more slender the plates, the higher the classification.
- The type of effect of action. Whereas the whole web is in compression under axial loading of the section, only half of it is in compression under pure bending loading and the susceptibility to local buckling is reduced in the later case.
- Material properties. In an elastic-perfectly plastic material;
  - If the Young's modulus is kept constant, higher yield strength means that the section has to be submitted to larger deformations before it develops the full plastic capacity. Sections with high yield strength are thus more prone to local buckling.
  - On the other hand, if the yield strength is kept constant, a lower Young's modulus means that the section has to be submitted to larger deformations before it develops the full plastic capacity. Sections with low Young's modulus are thus more prone to local buckling.

In fact, the parameter that drives the classification of the section with regard to the material properties is the square root between these two material properties, see Equation 5.4

$$\sqrt{E/f_y} \tag{5.4}$$

Because, at room temperature, the Young's modulus of steel can be regarded as a constant, the parameter that appears in the application rules is in fact the parameter ε given by Equation 5.5. Local buckling is most likely to occur for low values of this parameter.

$$\varepsilon = \sqrt{235/f_y} \tag{5.5}$$

where f_y is the yield strength of steel at room temperature, in N/mm^2.

The classification of a cross-section is made according to the highest class of its compression parts. Table 5.2 summarizes the limits of the width-to-thickness ratios (slenderness) for Class 1, 2 and 3, in case of internal compression parts (webs) and outstand flanges. Complete information can be found in EN 1993-1-1. For slenderness greater than the Class 3 limits, the cross-section should be taken as Class 4.

Table 5.2 Maximum slenderness for compression parts of cross-section

Class	Web		Flange compression
	Compression	Bending	
1	$\leq 33\varepsilon$	$\leq 72\varepsilon$	$\leq 9\varepsilon$
2	$\leq 38\varepsilon$	$\leq 83\varepsilon$	$\leq 10\varepsilon$
3	$\leq 42\varepsilon$	$\leq 124\varepsilon$	$\leq 14\varepsilon$

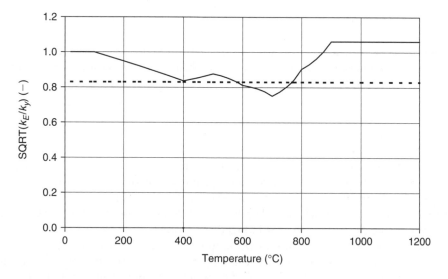

Fig. 5.4 Material property influencing local buckling

At elevated temperature, the Young's modulus as well as the yield strength are modified. The values at room temperatures are multiplied by $k_{E,\theta}$, respectively $k_{y,\theta}$, to give the values at elevated temperatures. If the material would remain elastic-perfectly plastic at elevated temperature, the parameter of Equation 5.4 would be transformed as indicated by Equation 5.6.

$$\sqrt{E_\theta / f_{y,\theta}} = \sqrt{\frac{k_{E,\theta}E}{k_{y,\theta}f_y}} = \sqrt{\frac{k_{E,\theta}}{k_{y,\theta}}}\sqrt{\frac{E}{f_y}} \tag{5.6}$$

The coefficients that describe the evolution of the Young's modulus and the yield strength, namely $k_{E,\theta}$ and $k_{y,\theta}$, follow two different functions of the temperature. The ratio $\sqrt{k_{E,\theta}/k_{y,\theta}}$ is thus also a function of the temperature, see Figure 5.4.

In Eurocode 3, the constant value of 0.85 has been considered for simplicity as an approximation of the function $\sqrt{k_{E,\theta}/k_{y,\theta}}$. Figure 5.4 shows that, for the practical temperature range from 500 to 800°C, this constant value is more or less an average value between all the possible values that can be calculated, namely 0.75 at 700°C and 0.90 at 800°C.

It has to be kept in mind that steel at elevated temperatures is not an elastic – perfectly plastic material and that the considerations based only on the Young's modulus and on the yield strength are only indicative.

The advantage of a constant value as opposed to a temperature dependent classification is that it prevents the following situation: for an infinitely small temperature increase in the range from 400 to 500°C or from 700 to 900°C, it could happen that a section has its classification improved, from Class 3 to Class 2 for example, and the steel member would, as a consequence, have its load bearing capacity increased by a temperature increase. This unrealistic result that would be created by the stepwise classification of the sections does not occur with a classification based on parameter ε that is not temperature dependent.

Finally, the classification of the sections in the fire situation is made according to the same rules as at ambient temperature but using the parameter defined by Equation 5.7 instead of Equation 5.5.

$$\varepsilon = 0.85\sqrt{235/f_y} \qquad (5.7)$$

This means that the whole classification process has to be made again in the fire situation, theoretically for each load combination because the classification depends on the effects of actions. The situation is complicated enough for the simple calculation models where the effects of actions are evaluated only once at time $t = 0$ but, for advanced calculation models, the classification should theoretically be done at every time step during the fire because of indirect fire actions.

In practice, some level of approximation must be tolerated. Generally, each section will be classified once for all in the fire situation depending on its most relevant load resisting mode. A beam will be classified as if acting in pure bending and a member that is essentially axially loaded will be classified as if acting in pure compression. It should be noted that, in moment resisting frames, a significant degree of bending is present, even in the columns.

5.6 How to calculate $R_{fi,d,t}$?

5.6.1 General principles

Generally speaking, the procedures used to calculate the design resistance of a steel member for the fire design situation are based on the same methods and equations as the ones used for the normal temperature situation, but modifying the mechanical properties of steel in order to take the temperature increase into account. This modification can be straightforward in the usual hypothesis of a uniform temperature in the section, somewhat more complex in case of a non-uniform temperature distribution.

This procedure is applicable only because the material model proposed by Eurocode 3 at elevated temperature does not contain an explicit creep term. Creep is deemed to be implicitly included in the stress strain relationship. As a consequence, the temperature leading to failure does not depend on the time required to reach this temperature and, hence, the thermal analysis and the mechanical analysis can be performed separately and in any order. For example, it is possible to determine first the critical temperature of a defined structure and then to choose the amount of thermal protection needed for this temperature not to be attained before a certain amount of time. This is possible only because the critical temperature is the same, whether it is reached within 20 minutes or within 2 hours. As a limit, Eurocode 3 states that this is valid provided that the heating rate in steel is comprised between 2 and 50°C/min., which is normally the case in building structures subjected to fire.

In fact, the procedures used to calculate $R_{fi,d,t}$ diverge in some aspects from the procedure used at room temperature to calculate R_d. This is the case namely;

(a) for the evaluation of the buckling length of continuous columns in braced frames,
(b) for the buckling and lateral torsional buckling curves,
(c) for the M-N interaction equations,
(d) for the classification of the sections,
(e) and for non uniform temperature distribution in beams.

The differences with the design at room temperature will be mentioned and discussed in the text wherever required. In fact, the differences from (b) to (d) can be traced down to the shape of the stress-strain diagram that is different at elevated temperature from the shape of the diagram that was considered when the design equations were established for room temperature conditions, see Figure 5.3. If the same equations are used at elevated temperature as at room temperature and simply f_y and E by the corresponding value at the elevated temperature, it is as if the material was following the path O-C-B (in a design equation based only on f_y) or the path O-D-B (in a design equation based on f_y and E) instead of the real path O-A-B. Some adaptations to the design equations established for room temperature conditions may thus be necessary when used at elevated temperatures.

The Eurocode proposes detailed equations for different types of effects of actions. These equations are presented and discussed in the next sections.

5.6.2 Tension members

In case of a uniform temperature, the equation proposed by Eurocode 3 for the design resistance of a tension member is Equation 5.8.

$$N_{fi,\theta,Rd} = k_{y,\theta} N_{R,d} [\gamma_{M,0}/\gamma_{M,fi}] \tag{5.8}$$

where:

$k_{y,\theta}$ is the reduction factor giving the effective yield strength of steel at temperature θ_a reached at time t,

N_{Rd} is the plastic design resistance of the cross-section for normal temperature design, according to EN 1993-1-1,

$\gamma_{M,0}$ is the partial safety factor for the resistance of cross-section at normal temperature,

$\gamma_{M,fi}$ is the partial safety factor for the relevant material property, for the fire situation.

Note: the recommended value for $\gamma_{M,0}$ and $\gamma_{M,fi}$ is 1.00, but different values may be defined in the National Annex.

It is as easy to use directly Equation 5.9 that is physically more meaningful.

$$N_{fi,\theta,Rd} = Ak_{y,\theta} [f_y/\gamma_{M,fi}] \tag{5.9}$$

where A is the cross-sectional area of the member.

It has to be realised that the utilisation of Equation 5.8 or Equation 5.9 means that the member must exhibit a stress-related strain of 2% for the full plastic load in tension to be mobilised. Added to a thermal strain in the order of magnitude of 1%, this means that the total elongation of the bar is near 3% at failure.

Eurocode 3, in 4.2.1 (5) and in Annex D, specifies that net-section failure at fastener holes need not be considered, provided that there is a fastener in each hole because, according to the Eurocode, the steel temperature is lower at connections due to the presence of additional material. It has yet been shown that this hypothesis is not safe in general, Franssen 2002. This is especially the case in protected members where the temperature is more likely to be nearly uniform or, in any case, after certain duration of a standard fire where the gas temperature tends to level off to a nearly constant level which, also, has a tendency to create a uniform situation in the steel structure. If the temperature distribution is uniform, there is no beneficial effect of added thermal massivity that can compensate for the reduction in net section.

The design resistance at time t of a tension member with a non-uniform temperature distribution across the cross-section may be determined by Equation 5.10 or Equation 5.11, with the latter equation leading to a conservative approximation.

$$N_{fi,t,Rd} = \sum_i A_i k_{y,\theta,i} [f_y/\gamma_{M,fi}] \tag{5.10}$$

where the subscript i refers to an elemental area of the cross-section in which the temperature is considered as uniform.

$$N_{fi,t,Rd} = Ak_{y,\theta max} [f_y/\gamma_{M,fi}] \tag{5.11}$$

where θ_{max} is the maximum temperature in the section at time t.

Application of Equation 5.10 makes sense only if the temperature distribution is symmetrically distributed in the section. If not, the mechanical centre of gravity in the section is moved by the non symmetrical variation of the yield strength and the section is submitted to tension and bending (for which there is no specific provisions in the Eurocode). In the case of a non symmetrical temperature distribution, it is preferable to accept an approximation and to use Equation 5.9 (uniform temperature distribution) or Equation 5.11 (maximum temperature in the section).

5.6.3 Compression members with Class 1, 2 or 3 cross-sections

This section is related to members that are submitted to axial compression; a separate section is indeed dedicated to members subject to combined bending and compression.

The following illustrates a scenario where the design in the fire situation differs somewhat from the design at normal temperature. The two differences are related to the evaluation of the buckling length and to the buckling curve that is used.

If the column is a continuous member that extends through several floors of a braced building and if each storey forms a separate fire compartment, then the buckling length of a column exposed to fire in an intermediate storey may be taken as $l_{fi} = 0.5\,L$ and in the top storey as $l_{fi} = 0.7\,L$ where L is the system length in the storey that is under fire.

The reason for considering these reduced lengths is that the stiffness of the column in the fire compartment decreases as its temperature increases, whereas the adjacent parts of the column that are located in the floors above or below remain at normal temperature and keep a constant stiffness. As a consequence, the adjacent parts become relatively stiffer and provide a significantly higher degree of restraint with respect to rotation. Therefore, the boundary conditions of the heated part of the column tend toward the condition of rotationally fixed supports, leading to the value of $0.5\,L$ (fixed-fixed supports) or $0.7\,L$ (fixed-free supports).

Although this is not explicitly mentioned in the Eurocode, it can be deduced that the buckling length of the column at the first floor should be equal to $0.5\,L$ or $0.7\,L$, depending on the boundary condition at the base of the column.

Numerical software that have been established for the analysis of structures at room temperature will not recognise this effect if the method utilised for calculating the slenderness of the members is based on the underlying hypothesis that the Young's modulus of the material is the same in every bar. An adaptation may be required if such software are used for analysing the structure in the fire situation.

The buckling curve of hot rolled sections subjected to fire has been studied in an ECSC research project (Schleich et al. 1988) and the results of this work have been incorporated in the Eurocode. The main results of this research work can also be found in Talamona et al. (1997) and in Franssen et al. (1998). The proposed equations have a form that is very similar to those proposed at normal temperature; the main differences are:

1. There are not any more several different buckling curves depending on the shape and dimensions of the cross-section or on the buckling axis, as was the case at room temperature.
2. The buckling curve now depends on the yield strength at room temperature, as was the case in some preliminary drafts of Eurocode 3 – Part 1, although this distinction had not been maintained in the final draft of the Eurocode for room temperature.

The successive steps to be followed to determine the design buckling resistance $N_{b,fi,t,Rd}$ of a member in compression with a uniform temperature θ_a are:

1. Determine the non-dimensional slenderness $(\overline{\lambda})$ based on material properties at room temperature, but using the buckling length in the fire situation as explained above, see Equation 5.12.

$$\overline{\lambda} = \frac{l_{fi}/\sqrt{I/A}}{\pi\,\sqrt{E/f_y}} \tag{5.12}$$

where:

I is the second moment of area of the cross-section

A is the area of the cross-section.

2. Determine the non-dimensional slenderness for the temperature θ_a $(\overline{\lambda}_\theta)$ according to Equation 5.13.

$$\overline{\lambda}_\theta = \overline{\lambda}\sqrt{k_{y,\theta}/k_{E,\theta}} \tag{5.13}$$

The term that multiplies the non-dimensional slenderness in Equation 5.13 is the invert of the term that is present in Equation 5.6 and presented on Figure 5.4. This reflects the fact that, because the Young's modulus decreases with temperature faster than the yield strength, the non-dimensional slenderness is higher at elevated temperatures, except for temperatures beyond 870°C, i.e. at temperatures that are not practically relevant.

3. Determine the imperfection factor of the utilised steel according to Equation 5.14

$$\alpha = 0.65\sqrt{235/f_y} \tag{5.14}$$

where f_y is the yield strength in N/mm^2.

4. Determine the coefficient φ_θ according to Equation 5.15.

$$\varphi_\theta = 0.5\left(1 + \alpha\overline{\lambda}_\theta + \overline{\lambda}_\theta^2\right) \tag{5.15}$$

5. Determine the buckling coefficient according to Equation 5.16.

$$\chi_{fi} = \frac{1}{\varphi_\theta + \sqrt{\varphi_\theta^2 - \overline{\lambda}_\theta^2}} \tag{5.16}$$

6. Determine the buckling resistance according to Equation 5.17.

$$N_{b,fi,\theta,Rd} = \chi_{fi}Ak_{y,\theta}f_y/\gamma_{M,fi} \tag{5.17}$$

Generally speaking, the above procedure has to be repeated twice, once for each buckling plane. In fact, it is sufficient to duplicate the first step and to pursue steps 2 to 6 for the plane that has the highest non-dimensional slenderness.

If the temperature distribution is non-uniform, the design fire resistance may be calculated according to the same procedure but on the basis of the maximum steel temperature. Yet, this is only admitted when designing using nominal fire exposure. It has indeed be shown, Anderberg 2002, that the lateral displacements that may be created by the non-uniform temperature can have a negative effect that outweighs the beneficial effect created by some parts of the section being colder than the maximum temperature. This is especially the case for slender members. Attention must be paid particularly to cantilevered column as encountered in fire resistant walls with no lateral support at the top. Thus, if the design is based on a realistic fire exposure, a similar degree of sophistication should be exercised in the mechanical analysis and these effects must be taken into account in a quantitative manner (utilising an advanced calculation model). This is clearly a case when, according to 2.4.2 (4) of Eurocode, "*the effects of thermal deformations resulting from thermal gradients across the cross-section need to be considered*". If, on the other hand, the fire exposure is represented by a nominal fire curve with its inherent arbitrary character, then Eurocode permits to make also an approximation in the mechanical analysis and to use the simple design equations based on the maximum temperature.

Although this is not explicitly mentioned in the Eurocode, it has to be recognised that a non-uniform temperature distribution that is symmetric in the section, for example the web of an I profile being hotter than the two flanges, does not produce any lateral displacement and it could be admitted in that case to use the simple design method, even if the fire exposure is not represented by a nominal curve. The restriction should apply to these non-symmetric temperature distribution that create lateral displacements, for example one of the flanges being colder than the other one.

Because the non-dimensional slenderness in the fire situation $\bar{\lambda}_\theta$ depends on the temperature, an iterative procedure appears if the critical temperature corresponding to a given applied load has to be determined (verification in the temperature or in the time domain, see Section 5.3). Convergence is usually very fast and one single iteration is usually sufficient if, for the first determined temperature, Equation 5.13 is approximated by Equation 5.18.

$$\bar{\lambda}_\theta = 1.2\bar{\lambda} \tag{5.18}$$

Application of the above equation leads to a first approximation of the critical temperature. The whole process can be repeated once with the exact Equation 5.13 being now used instead of Equation 5.18, with this first temperature being used to determine the non-dimensional slenderness. It will be observed that the second determined value for the temperature is not that much different from the first one and the iteration process need not to be continued.

The research that formed the basis of the proposed equation dealt mainly with hot rolled I sections.

It seemed quite logical to extend the results to welded I sections, probably because the influence of residual stresses that trigger a different behaviour at room temperature is less pronounced at elevated temperatures.

For sections with a totally different shape on the other hand, like for example angles or circular and rectangular hollow sections, utilisation of the proposed equation is indeed an extrapolation of the generated results to shapes that have not been considered in the study. This is the only alternative until further until further studies are undertaken on these types of section.

Numerical analysis tools that have been established for the analysis of columns at room temperature would thus not produce the appropriate result even if the appropriate buckling length is introduced, because the relative slenderness is now a function of the temperature and because a buckling curve that is specific to the fire situation must be used.

5.6.4 Beams with Class 1, 2 or 3 cross-section

It has first to be recognised that EN1993-1-2 does not give any definition of what a beam is. Because a separate section is dedicated to members under combined compression and bending, it can be concluded that a beam is a member under simple bending.

On the contrary to what is mentioned in the heading of this section, the Eurocode does not propose any method to design a beam; it simply gives a method to determine the resistance of *a section* in bending or in shear. The methods presented here by the authors are direct extrapolation of the methods used at room temperature.

5.6.4.1 Resistance in shear

The design shear resistance should be determined from Equation 5.19.

$$V_{fi,t,Rd} = k_{y,\theta,web} V_{RD}[\gamma_{M,0}/\gamma_{M,fi}] \tag{5.19}$$

where:

θ, web is the average temperature in the web of the section,
$k_{y,\theta,web}$ is the reduction factor for the yield strength of steel at the average temperature of the web,
V_{RD} is the shear resistance of the gross cross-section for normal temperature design according to EN 1993-1-1.

The fact that the average temperature of the web is mentioned here does not imply that the hypothesis of a uniform temperature is not admitted. Either a non-uniform distribution is considered, in which case the average temperature in the web is naturally considered for the shear resistance, or a uniform distribution is considered, in which case the average temperature in the web is equal to the uniform temperature in the section.

The method for designing a beam in shear is that Equation 5.20 is respected in any section of the beam.

$$V_{fi,Ed} \leq V_{fi,t,Rd} \tag{5.20}$$

where $V_{fi,Ed}$ is the shear force in the section in the fire design situation.

The shear force has to be determined by an elastic analysis of the effects of action in the beam if the cross-section is classified as Class 2 or Class 3, and by a plastic analysis in case of a Class 1 cross-section. Indeed, in the later case, the redistribution of plastic bending moment produced by the formation of plastic hinges also leads to a modification of the reaction forces and hence of the shear forces in the beam.

The shear force has also to be determined in order to take into account its effect on the bending resistance, see section 5.6.4.2 and 5.8.3.

5.6.4.2 Resistance in bending

5.6.4.2.1 Uniform temperature distribution
The design moment resistance of a section with a uniform temperature is given by Equation 5.21 proposed in the Eurocode or, equivalent, by Equation 5.22.

$$M_{fi,\theta,Rd} = k_{y,\theta}[\gamma_{M,0}/\gamma_{M,fi}]M_{Rd} \tag{5.21}$$

$$M_{fi,\theta,Rd} = k_{y,\theta}[f_y/\gamma_{M,fi}]W \tag{5.22}$$

where:

M_{Rd} is the plastic or elastic (function of the section classification) moment resistance of the gross cross-section for normal temperature design, allowing for the effects of shear if necessary, according to EN 1993-1-1,

W is the plastic modulus of the section W_{pl} for a Class 1 or a Class 2 section or the elastic modulus of the section W_{el} for a Class 3 section.

Note: Equation 5.21 and the comment that M_{Rd} has to be reduced for the effects of shear according to EN 1993-1-1 may lead to the conclusion that the ratio at room temperature $V_{Ed}/V_{pl,Rd}$ has to be considered in the reduction. In fact, Equation 5.22 shows that it is more consistent to consider the ratio at elevated temperature $V_{fi,d}/V_{fi,t,Rd}$.

The proposed method for designing a beam depends on the class of the section.

- If the cross-section is a Class 3 section, it has to be verified that the *elastically* determined bending moment in the fire design situation does not exceed the *elastic* design moment resistance in any section of the beam.
- If the cross-section is a Class 2 section, it has to be verified that the *elastically* determined bending moment in the fire design situation does not exceed the *plastic* design moment resistance in any section of the beam. In other words, the formation of one single plastic hinge is allowed, in the section where $M_{fi,Ed}$ is equal to $M_{fi,\theta,Rd}$ (plastic value), but no redistribution of bending moments is admitted.
- If the cross-section is a Class 1 section, a redistribution of bending moments may occur and lead to a plastic mechanism where the bending moment in the plastic hinges is determined by Equation 5.21 or 5.22 (plastic value). The resistance of the beam is the same as for a Class 2 section in a statically determinate beam, but is increased in a statically undeterminate beam.

5.6.4.2.2 Non-uniform temperature distribution
If the section is a Class 1 or a Class 2 section, it is possible to determine the plastic design resistance of the section ($M_{fi,t,Rd}$) taking into account the value of the yield strength in each part of the section. The Eurocode specifies the following equation to compute $M_{fi,t,Rd}$.

$$M_{fi,t,Rd} = \sum_i A_i z_i k_{y,\theta,i} f_{y,i}/\gamma_{M,fi} \tag{5.23}$$

where:

A_i is the cross section of an elemental area of the cross-section with a temperature θ_I,

z_i is the distance from the plastic neutral axis to the centroid of the elemental area A_i.

The position of the plastic neutral axis changes continuously during the course of the fire and has to be determined at the relevant moment in time. This position can be determined from the fact that the plastic resistance on one side of the axis is equal to the plastic resistance on the other side of the axis.

There is a sentence in Clause 4.2.3.3 (2) of EN 1993-1-2 that says, about Equation 5.23, that f_y must be *"taken as positive on the compression side of the plastic neutral axis and negative on the tension side"*. This sentence is without any merit and must be ignored. It would only make sense to take the yield strength as negative on the tension side if the distance z_i would also be taken as negative on the tension side, which is physically not correct because a distance is always positive. It makes sense to count some elemental areas as positive and some others as negative in Equation 5.24 if it is used in order to determine the position of the plastic neutral axis (this equation is valid if the yield strength is uniform in the section).

$$\sum_i A_i k_{y,\theta,i} = 0 \tag{5.24}$$

More important is the fact to recognise that if an elemental area comprises the neutral axis, it must then be divided into two sub-areas, one above and one below the neutral axis. This is the case, for example, in the web of an I section or in the webs of a rectangular hollow structural section.

There is in the Eurocode no equation similar to Equation 5.23 for Class 3 sections. Nothing would theoretically speak against the fact to determine the location of the elastic neutral axis, according to equation 5.25 for example.

$$\sum_i A_i(y_i - y)k_{E,\theta,i} = 0 \tag{5.25}$$

where:

y_i is the co-ordinate in an arbitrary reference axis of the centroid of the elemental area A_i,

y is the co-ordinate in the same reference axis of the elastic neutral axis of the section,

$k_{E,\theta,i}$ is the reduction factor of the Young's modulus of steel at temperature θ_i.

The elastic stiffness in bending could then be determined according to Equation 5.26.

$$EI_{el,t} = \sum_i A_i z_i k_{E,\theta,i} E \tag{5.26}$$

where z_i would now be the distance from the elastic neutral axis.

The stress should then be checked against the yield strength in each area according to Equation 5.27.

$$\frac{M_{Ed,fi} z_i}{EI_{el,t}} \leq k_{y,\theta,i} f_y \tag{5.27}$$

Yet, the procedure described by Equation 5.25 to 5.27 is not proposed in the Eurocode for Class 3 sections. The more simple procedure described hereafter is proposed in EN 1993-1-2.

In a member with non uniform temperature distribution, the design bending moment resistance of a cross-section with non-uniform cross-section may be determined from Equation 5.28 for Class 1 or Class 2 sections (in which case this is an alternative to Eq. 5.23) or from Equation 5.29 for Class 3 sections (in which case this is the standard procedure).

$$M_{fi,t,Rd} = k_{y,\theta}[f_y/\gamma_{M,fi}]\frac{W_{pl}}{\kappa_1\kappa_2} \tag{5.28}$$

$$M_{fi,t,Rd} = k_{y,\theta,\max}[f_y/\gamma_{M,fi}]\frac{W_{el}}{\kappa_1\kappa_2} \tag{5.29}$$

where:

θ in Equation 5.28 is for Class 1 and 2 sections, a uniform steel temperature in the section that is not thermally influenced by the supports,

θ, max in Equation 5.29 is for Class 3 sections, the maximum steel temperature reached at time t,

κ_1 is an adaptation factor for non-uniform temperature in the cross-section,

κ_2 is an adaptation factor for non-uniform temperature along the beam.

The following factors should be given due consideration while using Equations 5.28 and 5.29.

- Equations 5.28 and 5.29 have been developed for the simplest case when there is no reduction of the bending resistance due to the effects of shear. The effect of shear on bending moments has to be taken into account if necessary.
- Equations 5.28 and 5.29 take into account the fact that the temperature in steel members may be colder (at least in the heating phase, for real fires) in the zones near the support than in the zones that are far away from the supports, at mid span for example. This is because the material that physically constitutes the support of a beam may shield the beam locally from the fire exposure and may act as heat sink. This is the case, for example, if the support is a masonry or a concrete wall. It has indeed been observed after real fires or in experimental tests that the plastic hinge leading to failure was displaced from the support toward the centre of the span by a distance ranging from 20 to 100 cm.

  Yet, the real temperature distribution near the supports can not be determined precisely and a simple method only allows determination of the temperature in the central zones of the beam. Because the structural analysis will be based on this undisturbed temperature distribution, a correction factor has been introduced, namely the factor κ_2. The Eurocode says that this factor must be given the value of 0.85 at the supports of statically indeterminate beams and 1.0 in all other cases. The probable reason might be the Eurocode considering a statically determinate beam as a beam simply supported on two end supports. In that case, even if the temperature may be colder near the supports, this has no effect on the fire resistance because the bending moment at supports is close to zero. In a continuous beam, on the other hand, the cold effect may be significant if it occurs in the intermediate supports where the bending moment has the highest values. Theoretically speaking, the same effect could also exist at the single support of a cantilever beam which is also a statically determinate beam, but the Eurocode does not allow to take the effect into account in that case, perhaps because the formation of a single plastic hinge in a statically indeterminate beam does not lead immediately to failure whereas one single hinge leads to failure in a statically determinate beam and more caution has to be exercised. By analogy, the authors of this book recommend that the effect not be taken into account in the most exterior supports of a continuous beam with a cantilever part because one single hinge at that location leads to the immediate failure of the cantilever.

The authors also believe that the permission to use the value of $\kappa_2 = 0.85$ is not automatically granted at every support of every statically indeterminate beam; the designer must be convinced that the temperature at that location is really lower than in the central parts of the beam and the rational for this should be justified. Such an effect, for example, would certainly not be encountered if the intermediate supports of the beam are steel tension rods or axially loaded steel columns with a thermal massivity that is smaller than the thermal massivity of the beam, which means that the temperature might actually be somewhat higher at the supports than in the span.

It has to be noted that this effect of colder zones near the supports is not systematically taken into account if the analysis of a continuous steel beam is performed by the advanced calculation model. Indeed, taking it into account correctly would require a 3D thermal modelling of the zone near the supports. It would be possible to introduce a small length of the beam near the support that is prevented from the effect of the fire, but this would be an approximation arbitrarily introduced. This effect thus generally not considered by the advanced calculation model with the consequence that the simple model can yield results that are on the unsafe side compared to the more advanced model.

• Equations 5.28 and 5.29 take into account the fact that the temperatures in the section of a beam that supports a concrete slab are somewhat lower than the temperatures that are calculated by the simple method. Indeed, the simple method allows taking into account the fact that the upper side of the top flange is not submitted to the fire, by a simple modification of the thermal massivity. What is not taken into account by the simple model is the heat sink effect, the fact that some heat is transferred from the top flange of the section to the concrete deck, which delays the temperature increase in the steel section. In order to take this beneficial effect into account, the factor κ_1 is given the value of 0.70 for unprotected beams and 0.85 for protected beams exposed on three sides, with a composite or concrete slab on side four.

Although this is not explicitly mentioned in Eurocode 3, the authors believe that, in order to be consistent with Eurocode 4, a steel beam supporting a composite slab should be considered as heated on 3 sides only if the area of the upper flange that is covered with the corrugated steel profile of the composite slab is at least equal to 90% of the whole area of the upper flange; if not, the steel section should be considered as heated on four sides and the value of the factor κ_1 kept to 1.0.

Precise evaluation of the plastic capacity of a steel section covered with a concrete slab, for example by means of an advanced calculation model taking the heat sink effect into account, fails by far to show such a huge increase compared to the plastic capacity that can be calculated when the heat sink is not taken into account. Burgess et al. (1991) have shown that this beneficial effect is about 7% in a symmetrical section. In fact, the value of 0.70 for unprotected sections (i.e. an increase of $(1 - 0.7)/0.7 = 43\%$) has been considered by the draft team on EN 1993-1-2 on the basis of results from an experimental test series performed in the U.K. (Wainman & Kirby 1988). In fact, the test reports mention that "*thin gauge steel reinforcing tang*" had been welded on the top flange of the sections and cast into the concrete of the supported slabs and may have thus produced some level of composite action. The beneficial effect accounting for the heat sink effect was reduced by a factor of 2 in the Eurocode, from 0.70 to 0.85, for protected beams in order to reflect the fact that the advanced numerical calculations show an even less pronounced effect in that case.

As long as the temperature at mid span does not reach 560°C, the reduction factor of the effective yield strength $k_{y,\theta}$ is higher than 0.595. Strict application of Equation 5.28 or 5.29, in which $\kappa_1\kappa_2 = 0.70 \times 0.85 = 0.595$, would then yield a resistance to bending at the support that is higher at elevated temperatures than at room temperature! It may be wise to limit $M_{fi,t,Rd}$ to M_{Rd}.

- The temperature that has to be taken into account to evaluate the bending resistance of the section is not exactly the same in Equation 5.28 and in Equation 5.29.
    - For Class 1 and Class 2 sections, see Equation 5.28, the temperature is the uniform temperature calculated in the central part of the beam, far away from the effect of the supports.
    - For Class 3 sections, see Equation 5.29, the temperature is "*the maximum steel temperature reached at time t*", and it is not straightforward to know which temperature exactly has to be taken into account.

It seems obvious to take also this temperature in a section that is not thermally influenced by the supports because; firstly, it is not possible to calculate the temperature on the supports; secondly, there would be no reason to use a κ_2 factor if the temperature would be taken in the region of the supports; thirdly, the definition of this temperature in Clause 4.2.3.4 (2) is finished by a comment that reads "*see 3*", which might be understood as "*see 4.2.3.3*", i.e. far away from the supports as for Class 1 and Class 2 sections.

Where in the section is the maximum temperature? The opinion of the authors is that the designer may consider the temperature in the section as uniform if the section is exposed on four sides, in which case only the factor κ_2 would be considered ($\kappa_1 = 1.0$). If the beam supports a concrete slab, it is not easily determined a priori where the maximum temperature is, but it seems certain that this is not in the top flange. One may argue that the maximum temperature is in the web because this is the thinnest plate of the section, but one could also argue that, on the contrary, the web could feel the cooling influence from the top flange, especially if the section is no very deep. In fact, the most practical way is to calculate the temperature as if the section was exposed on four sides, i.e. on the basis of

Table 5.3 Parameters for beam design

Section Class	Exposed on all four sides	Exposed on three sides with slab on side four
1, 2	$\kappa_1 = 1.0$ θ_a computed considering A/V for four sides	$\kappa_1 = 0.7$ θ_a computed considering A/V for three sides
3	$\kappa_1 = 1.0$ $\theta_{a,max}$ computed considering A/V for four sides	$\kappa_1 = 0.7$ $\theta_{a,max}$ computed considering A/V for four sides

the massivity factor of the section exposed on four sides. This is in fact the temperature that would be calculated if the influence of the concrete deck was totally ignored. This temperature may also be considered to correspond, more or less, to the average temperature in the lower half of the section.

Table 5.3 summarizes the above considerations and presents the parameters to be considered for the beam design, function of the cross-section class and exposure.

The proposed method for designing a beam in the case of non uniform temperature distribution is the same as the one proposed in 5.6.4.2.1 for uniform temperature distribution. The bending capacity of the section still depends on the class of the section. The difference is that the plastic capacity may not be the same at the supports and in the spans because of the κ_2 correction factor taking into account the effect of the supports.

5.6.4.3 Resistance to lateral torsional buckling

The design lateral torsional buckling resistance moment of a beam ($M_{b,fi,t,Rd}$) should be determined according to Equation 5.30.

$$M_{b,fi,t,Rd} = \chi_{LT,fi} W_y k_{y,\theta,com} f_y / \gamma_{M,fi} \qquad (5.30)$$

where:

W_y is the plastic modulus of the section ($W_{pl,y}$) for Class 1 or 2 sections, the elastic modulus of the section ($W_{el,y}$) for Class 3 sections,

$k_{y,\theta,com}$ is the reduction factor for the yield strength of steel at the maximum temperature in the compression flange,

$\chi_{LT,fi}$ is the coefficient for lateral torsional buckling, calculated from Equation 5.31.

$$\chi_{LT,fi} = \frac{1}{\varphi_{LT,\theta,com} + \sqrt{\varphi^2_{LT,\theta,com} - \bar{\lambda}^2_{LT,\theta,com}}} \qquad (5.31)$$

with

$$\varphi_{LT,\theta,com} = 0.5(1 + \alpha\bar{\lambda}_{LT,\theta,com} + \bar{\lambda}^2_{LT,\theta,com}) \tag{5.32}$$

$$\alpha = 0.65\sqrt{235/f_y} \tag{5.33}$$

$$\bar{\lambda}_{LT,\theta,com} = \sqrt{k_{y,\theta,com}/k_{E,\theta,com}}\,\bar{\lambda}_{LT} \tag{5.34}$$

The temperature considered for evaluating $k_{y,\theta}$ and $\bar{\lambda}_{LT,\theta}$ can conservatively be taken as the uniform temperature in Class 1 and 2 sections and as the maximum temperature for Class 3 sections. The authors recommend calculating this temperature according to Table 5.3.

While designing a beam, considering the lateral torsional buckling, the following equation is to be satisfied:

$$M_{fi,Ed,max} \leq M_{b,fi,t,Rd}$$

where $M_{fi,Ed,max}$ is the maximum bending moment on the beam between two lateral restraints, in the fire design situation.

Note: EN 1993-1-2 does not take into account the moment distribution between lateral restraints of the beam in the computation of $M_{b,fi,t,Rd}$, which means that a uniform distribution of the maximum moment along the beam is considered, normally leading to conservative result. On the contrary, this aspect is considered in EN 1993-1-1 through a factor f that increases the resisting moment, computed function of the shape of bending moment diagram. By means of numerical investigations (Vila Real et al. 2004, 2005 and Lopes et al. 2004) showed that the shape of bending moment diagram along the beam is important also in case of fire and proposed a formula for the factor f to be applied in the fire design situation, as is the case at room temperature.

5.6.5 Members with Class 1, 2 or 3 cross-sections, subject to combined bending and axial compression

This section is related to members that are submitted to axial compression and to bending. This is a very common situation, for example, in moment resisting frames. The design resistance of such a member subjected to combined bending and axial compression and a uniform temperature is verified according to Equation 5.35 and 5.36.

$$\frac{N_{fi,Ed}}{\chi_{min,fi}Ak_{y,\theta}\frac{f_y}{\gamma_{M,fi}}} + \frac{k_y M_{y,fi,Ed}}{W_y k_{y,\theta}\frac{f_y}{\gamma_{M,fi}}} + \frac{k_z M_{z,fi,Ed}}{W_z k_{y,\theta}\frac{f_y}{\gamma_{M,fi}}} \leq 1 \tag{5.34}$$

$$\frac{N_{fi,Ed}}{\chi_{z,fi}Ak_{y,\theta}\frac{f_y}{\gamma_{M,fi}}} + \frac{k_{LT} M_{y,fi,Ed}}{\chi_{LT,fi}\,W_y k_{y,\theta}\frac{f_y}{\gamma_{M,fi}}} + \frac{k_z M_{z,fi,Ed}}{W_z k_{y,\theta}\frac{f_y}{\gamma_{M,fi}}} \leq 1 \tag{5.35}$$

where:

W_y, W_z are the plastic modulus $W_{pl,y}$, $W_{pl,z}$ for Class 1 and 2 sections and the elastic modulus $W_{el,y}$, $W_{el,z}$ for Class 3 sections.

$\chi_{min,fi}$ is the smaller of $\chi_{y,fi}$ and $\chi_{z,fi}$, these being calculated from Equation 5.16,

$\chi_{LT,fi}$ is calculated from Equation 5.31,

$$k_{LT} = 1 - \frac{\mu_{LT} N_{fi,Ed}}{\chi_{z,fi} A k_{y,\theta} \frac{f_y}{\gamma_{M,fi}}} \leq 1 \tag{5.36}$$

with $\mu_{LT} = 0,15 \,\overline{\lambda}_{z,\theta} \beta_{M.LT} - 0,15 \leq 0,9$ \hfill (5.37)

$$k_y = 1 - \frac{\mu_y N_{fi,Ed}}{\chi_{y,fi} A k_{y,\theta} \frac{f_y}{\gamma_{M,fi}}} \leq 3 \tag{5.38}$$

with: $\mu_y = (1,2\beta_{M.y} - 3)\overline{\lambda}_{y,\theta} + 0,44\beta_{M.y} - 0,29 \leq 0,8$ \hfill (5.39)

$$k_z = 1 - \frac{\mu_z N_{fi,Ed}}{\chi_{z,fi} A k_{y,\theta} \frac{f_y}{\gamma_{M,fi}}} \leq 3 \tag{5.40}$$

with: $\mu_z = (2\beta_{M.z} - 5)\overline{\lambda}_{z,\theta} + 0,44\beta_{M.z} - 0,29 \leq 0,8$ and $\overline{\lambda}_{z,\theta} \leq 1,1$ \hfill (5.41)

β_M, the equivalent uniform moment factor, is defined in Figure 5.6 taken from Eurocode 3.

The following points should be kept in mind in the application of above Equations.

- Some confusion may arise from the fact that, in Equation 5.34 to 5.41, the subscripts y and z normally refer to the two main axes of the section, except in $k_{y,\theta}$ where y refers to f_y.
- It must be mentioned that, in some drafts of prEN1993-1-2, equations 5.38 and 5.40 are not correctly reproduced. An editorial error led to the erroneous suppression of the factors μ_y in 5.38 and μ_z in 5.40.
- The shape of the proposed equations is similar to the shape of the equations that were present for room temperature in ENV 1993-1-1 and, hence, is different from the shape of the equations now proposed for room temperature in EN 1993-1-1. The reasons behind this inconsistency are discussed hereafter and illustrated in Figure 5.5.

In fact, the research work carried out in the early 90's for deriving buckling curves and M-N interaction equations for steel sections subjected to fire have taken as the basis Eurocode provisions existing at that time, i.e. ENV 1993-1-1. The equations proposed for elevated temperatures in ENV 1993-1-2 (1995) are thus similar to the ENV equations at room temperature published at that time. When the draft team of EN 1993-1-2 was at work, its members were aware of the fact that, at the same time, the draft team of EN 1993-1-1 was working on the interaction equations at room

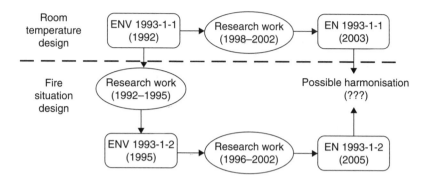

Fig. 5.5 Evolution of different Eurocodes

temperature, with the aim of proposing an improved model for ambient conditions. However the draft team decided to adhere for the fire situation to the equations proposed in ENV 1993-1-2 and to take them directly on board in EN 1993-1-2, with a new improvement related to lateral torsional buckling. The two main reasons were:

(a) the equation of ENV 1993-1-2 had been validated and calibrated in the fire situation, whereas the new equations introduced in EN 1993-1-1 had never been validated in the fire situation

(b) two different proposals were under consideration by draft team of EN 1993-1-1 and it was impossible for the members of EN 1993-1-2 to know which one would finally be chosen for ambient situation.

As a consequence, in the current version of Eurocode 3, there is no similarity between the M-N interaction curves at room temperature and in the fire situation. Recent published work seem to indicate that it could be possible to restore the similarity and that the interaction curves that are now present in the cold Eurocode could be adapted for the fire situation, taking into account the main results of the scientific research made in the 90's for the fire situation (Vila Real et al. 2003).

- As already mentioned in Section 5.6.3, the research that is the basis of the proposed equations dealt mainly with hot rolled I sections. For sections with a totally different shape, like for example angles or circular and rectangular hollow sections, utilisation of the proposed equation is indeed an extrapolation of the obtained results to shapes that have not been covered in the original studies. If the section has no weak axis, lateral torsional buckling is not possible and Equation 5.35 does not apply; only Equation 5.34 has to be used.

- Equations 5.34 and 5.35 in fact check the stability of a member, but the resistance of the section is not checked. It has yet to be mentioned that the proposed equations have been derived on the base of a comprehensive set of numerical simulations by the advanced calculation model. In these finite element analyses, the resistance of the sections was automatically taken into account and, if extensive yielding at one extremity of the member occurred in case, for example, of a bi-triangular bending

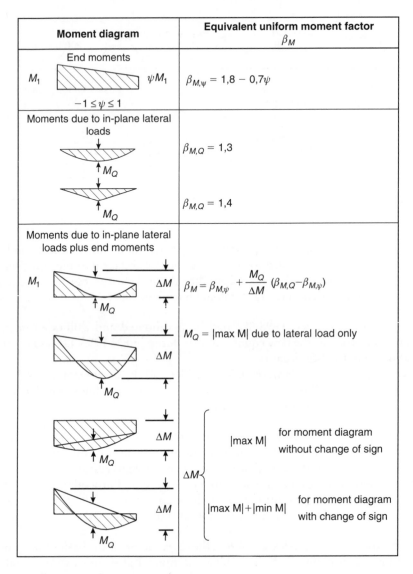

Moment diagram	Equivalent uniform moment factor β_M
End moments M_1 $\quad$ ψM_1 $-1 \leq \psi \leq 1$	$\beta_{M,\psi} = 1{,}8 - 0{,}7\psi$
Moments due to in-plane lateral loads M_Q M_Q	$\beta_{M,Q} = 1{,}3$ $\beta_{M,Q} = 1{,}4$
Moments due to in-plane lateral loads plus end moments M_1 $\quad$ M_Q $\quad$ ΔM M_Q $\quad$ ΔM M_Q $\quad$ ΔM M_Q $\quad$ ΔM	$\beta_M = \beta_{M,\psi} + \dfrac{M_Q}{\Delta M}(\beta_{M,Q} - \beta_{M,\psi})$ $M_Q = |\max M|$ due to lateral load only $\Delta M \begin{cases} |\max M| & \text{for moment diagram without change of sign} \\ |\max M| + |\min M| & \text{for moment diagram with change of sign} \end{cases}$

Fig. 5.6 Equivalent uniform moment factor

moment diagram, this was automatically reflected by a loss of equilibrium. Thus it can then be inferred that these equations for the stability of the member in fact include verification of equilibrium in the sections.

• It has to be noted that Eurocode 3 is not fully consistent with regard to consideration of the moment distribution in the element. If a uniform distribution of the maximum moment is considered in the verification of the beam resistance to lateral torsional buckling, see Section 5.6.4.3, the shape of the bending moment diagram is considered in case of an element subjected to combined bending and axial compression.

- According to the original research work (Talamona 2005, and Schleich et al. 1998) the correct expressions for μ_y and μ_z are:

$$\mu_y = (2\beta_{M,y} - 5)\,\overline{\lambda}_{y,\theta} + 0,44\beta_{M,y} + 0,29 \leq 0,8 \quad \text{with } \overline{\lambda}_y \leq 1,1$$

and

$$\mu_z = (1,2\beta_{M,z} - 3)\overline{\lambda}_{z,\theta} + 0,71\beta_{M,z} - 0,29 \leq 0,8$$

5.6.6 Members with Class 4 cross-sections

Tension members can of course develop the full plastic capacity of the section because there is no possibility of local buckling of a plate in tension. These members are designed according to the provisions of Section 5.6.2.

For members submitted to any other effect of action, it may be assumed that the load bearing function of the member is ensured as long as the steel temperature in all sections does not exceed a critical temperature, the recommended value of which is 350°C (to be chosen as a nationally determined parameter).

This is indeed a very simple design method, but a very restrictive one. In fact, Class 4 sections are those that have the thinnest plates and, hence, those that exhibit the fastest heating rates. Except when effective thermal protection is applied, these sections attain critical temperature in a very short time and usually fail to meet to required fire resistance time.

A first possibility to go beyond this critical temperature failure criterion could be to accept an elastic design based on the appropriate reductions of the yield strength of steel, provided that all parts of the section are shown to be in tension. Such a stress distribution would in fact be in equilibrium for members submitted to tension and to a small amount of bending. In other words, the eccentricity of the applied axial force is small.

For all other situations, Eurocode 3 allows a more precise determination of the fire resistance time explained in Annex E. In fact, the text of Eurocode 3 says "*For further information see Annex E*". Annex E gives more than just information about the concept of critical temperature. It gives in fact a simple calculation model to be used for Class 4 cross-sections.

This model is based on three basic concepts (Ranby 1998), namely:

1. The same equations as those presented in sections 5.6.3 for members in compression, in 5.6.4 for beams and in 5.6.5 for combined bending and compression, are used.
2. In these equations, the area is replaced by the effective area and the section modulus by the effective section modulus in order to take local buckling into account. These effective properties are based on the material properties of steel at 20°C. This can lead to a curious situation. Indeed, the classification is more severe in the fire situation than at room temperature, see section 5.5 and Equation 5.7, and it could occur that a cross-section is classified as Class 4 in the fire situation whereas it was classified as Class 3 at room temperature. In that case, the effective properties are equal to the basic properties (no reduction) because the effective

properties are based on the properties at room temperature, i.e. for a Class 3 section.

3. In these equations, the design strength of steel should be taken as the 0.2 percent proof strength, for the resistance to compression, to shear, as well as to tension. This evolution of the 0.2 percent proof yield strength is given as a function of temperature in Annex E. It can be seen that the reduction is nearly the same as the reduction exhibited by the Young's modulus.

5.7 Design in the temperature domain

Eurocode 3 dedicates a separate section, namely Section 4.2.4, to the design in the temperature domain. The basic idea is to obtain directly the critical temperature from the load level or, as called in the Eurocode, the degree of utilisation.

For all members in tension and for all members with Class 1, 2 or 3 section, the degree of utilisation is obtained from Equation 5.42.

$$\mu_0 = E_{fi,d}/R_{fi,d,0} \tag{5.42}$$

where:

$E_{fi,d}$ is the design effect of actions in case of fire,

$R_{fi,d,0}$ is the design resistance of the member in the fire situation, i.e. determined with the equations mentioned here in Section 5.6, but at time $t = 0$, i.e. for a temperature of 20°C.

The critical temperature is then given by Equation 5.43 as a function of the degree of utilisation.

$$\theta_{a,cr} = 482 + 39.29 \ln\left[\frac{1}{0.9674\,\mu_0^{3.833}} - 1\right] \tag{5.43}$$

In fact, working directly in the temperature domain with Equation 5.42 and 5.43 is only valid if the design resistance in case of fire $R_{fi,d,t}$ is directly proportional to $f_y(\theta)$, see Equation 5.44.

$$R_{fi,d,t} = m f_y(\theta) \tag{5.44}$$

with m a constant.

Indeed, in that case, the basic design equation can be transformed as follows:

$$\begin{aligned} E_{fi,d} \le R_{fi,d,t} &= m f_y(\theta) \\ &= m k_{y,\theta} f_y \\ &= R_{fi,d,0} k_{y,\theta} \end{aligned} \tag{5.45}$$

$$\begin{aligned} k_{y,\theta} &\ge E_{fi,d}/R_{fi,d,0} = \mu_0 \\ \theta &= k_{y,\theta}^{-1}(k_{y,\theta}) = k_{y,\theta}^{-1}(\mu_0) \end{aligned} \tag{5.46}$$

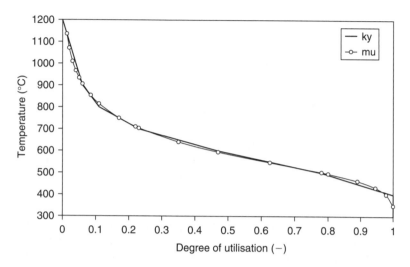

Fig. 5.7 Comparison between inverse of yield strength and degree of utilisation factor

This shows that Equation 5.43 that gives the critical temperature as a function of the degree of utilisation must be the invert of the function that gives the reduction of the effective yield strength as a function of the temperature.

Figure 5.7 shows the comparison between the invert of $k_{y,\theta}$ as taken from Table 3.1 of the Eurocode and Equation 5.43. It shows that the two curves are very close for the domain of application mentioned in the Eurocode, i.e. for degrees of utilisation not below 0.013, i.e. for temperatures not higher than 1135°C.

A design should thus yield very similar results, be it carried out in the load domain or in the temperature domain. It is not forbidden, instead of Equation 5.43, to use exactly the invert of the function $k_{y,\theta}$ when working in the temperature domain, i.e. to interpolate in Table 3.1 of the Eurocode in order to determine θ_{cr} as a function of μ_0 although the heading of the Table gives θ as a function of $k_{y,\theta}$. In that case, the design in the temperature domain will give exactly the same result as a design in the load domain.

Similar, or exactly equal results will be obtained only if the load bearing capacity in the fire situation is strictly proportional to the effective yield strength, see Equation 5.44. If, on the contrary, the evolution of the Young's modulus also plays a role in the evolution of $R_{fi,d,t}$, it is then not possible anymore to apply directly Equation 5.43. The critical temperature in such a scenario can be determined only by successive verifications in the load domain. This is the case for example for columns under buckling, for members under combined bending and compression, for members under lateral torsional buckling and for Class 4 sections. This is explained in Clause 4.2.4 (2) of the Eurocode with a statement: "*Except when considering deformation criteria or when stability phenomena have to be taken into account*". This should be interpreted as "*Except when the Young's modulus plays a role in the fire resistance*". As can be seen from the list given above, application of Equation 5.43 is restricted to a rather limited number of cases such as, for example, for bending in Class 1, 2 or 3 sections and for members in tension.

Another situation that does not allow direct application of Equation 5.43 to find directly the critical temperature is when the bending resistance at the supports of statically indeterminate beams has to be reduced for the effects of shear, see Section 5.8.3.

Although is it not very clear from the text, the note under Table 4.1 in the Eurocode, "*The national annex may give default values for critical temperatures*", is mainly introduced to allow member states to introduce some values of critical temperatures for various situations, for example 540°C in beams and 520°C when buckling may play a critical role. The designer would then have the possibility not to perform any structural analysis at all in the fire situation provided that enough thermal insulation is applied in order to limit the temperature increase at the required fire resistance time to be below the prescribed critical temperature.

This option of fixed critical temperatures is quite convenient for the designer. Apart from the fact that theoretical validation for the prescribed critical temperatures is rather weak, this option usually yields to a quite uneconomic design because it amounts to neglect any favourable effect that may arise from a lower degree of utilisation.

5.8 Design examples

For the examples proposed in this section, a default value of 1.0, as proposed in the Eurocode, has been adopted for the partial safety factor $\gamma_{M,fi}$.

5.8.1 *Member in tension*

A member with a circular section, diameter $D = 250$ mm, thickness $d = 5$ mm, yield strength $f_y = 355$ N/mm^2, is subjected to an axial tension load in the fire situation $E_{d,fi} = 100$ kN. The required fire resistance time is $t_{req} = 30$ minutes.

Perform the verification in the load domain, in the time domain and in the temperature domain.

5.8.1.1 *Verification in the load domain*

For the required fire resistance time $t_{req} = 30$ minutes, the design resistance of the tension member must be higher than the design load, Equation 5.2.

Perimeter of the section: $A_m = \pi D = \pi 0.25 = 0.785$ m
Area of the section: $V = \pi(D^2 - (D - 2d)^2)/4 = \pi\,(0.25^2 - 0.24^2)/4 = 0.003848$ m^2.
Massivity factor: $A_m/V = 0.785/0.003848 = 204$ m^{-1}
Note: this value is very close to $1/d = 200\,m^{-1}$

Interpolation in Table I.1 (see Annex I) between 200 and 400 m^{-1} yields a temperature of 828°C after 30 min.

Interpolation in Table II.2 (see Annex II) between 800 and 900°C yields $k_{y,\theta} = 0.096$ at 828°C.
$f_{y,\theta} = k_{y,\theta} f_y = 0.096 \times 355 = 34.08$ N/mm^2

$N_{fi,\theta,Rd} = Vf_{y,\theta} = 3848\,\text{mm}^2 \times 34.08\,\text{N/mm}^2 = 131\,\text{kN} > N_{fi,Ed}$ (100 kN)

The safety margin in terms of applied load is $(131 - 100)/100 = 31\%$

5.8.1.2 Verification in the time domain

The fire resistance time $t_{fi,d}$ of the tension member must be higher than the required fire resistance. It is reached when $N_{fi,\theta,Rd} = N_{fiE,d}$

Failure occurs when $N_{fi,\theta,Rd} = Vf_{y,\theta} = 3848 \times f_{y,\theta} = 100\,\text{kN} \Rightarrow f_{y,\theta} = 25.99\,\text{N/mm}^2$
$k_{y,\theta} = f_{y,\theta}/f_y = 25.99/355 = 0.0732$.

Interpolation in Table II.2 (see Annex II) between 0.06 and 0.11 yields a failure temperature of 874°C for a reduction factor of 0.0732.

Table I.1 (see Annex I) gives a time of 39 minutes to reach a temperature of 874°C in a section with a massivity factor of 200^{-1} ($\approx 204\,\text{m}^{-1}$).
Note: See Section 5.8.1.1 for the calculation of the massivity factor.
The safety margin in terms of time is 9 minutes.

5.8.1.3 Verification in the temperature domain

$$R_{d,fi,0} = Vf_y = 3848 \times 355 = 1366\,\text{kN}$$

$$\mu_0 = E_{fi,d}/R_{fi,d,0} = 100/1366 = 0.0732$$

$$\theta_{cr} = 39.19\ln\left[\frac{1}{0.9674\mu_0^{3.833}} - 1\right] + 482$$

$$= 39.19\ln\left[\frac{1}{0.9674 \times 0.0732^{3.833}} - 1\right] + 482$$

$$= 876°C$$

For the required fire resistance time $t_{req} = 30$ minutes and for a massivity factor $A_m/V = 204\,\text{m}^{-1}$, $\theta_d = 828°C$, see Section 5.8.1.1.
The safety margin in terms of temperature is 48°C.

Notes:

1. *It can be observed that the degree of utilization μ_0 used in the temperature domain, 0.0732, is in fact equal to the reduction factor for the yield strength $k_{y,\theta}$ corresponding to the failure time $t_{fi,d}$ determined in the time domain.*
2. *The critical temperature $\theta_{cr} = 876°C$ determined in the temperature domain differs by less than 0.3% from the temperature $\theta_d = 874°C$, corresponding to the failure of the element determined in the time domain. The reason of this difference is explained in Section 5.7, see Figure 5.7.*

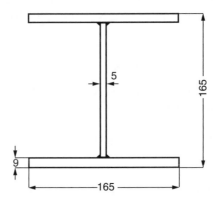

Fig. 5.8 Section of the column under axial compression

5.8.2   *Column under axial compression*

A simply supported column with **I** welded cross-section (see Figure 5.8) and length of 2,90 m is made of steel with $f_y = 235 \, \text{N/mm}^2$. The column is supporting a design load of 410 kN in fire situation and is exposed to fire on all sides.

1. What is the fire resistance time of the column to the ISO834 standard fire, considering buckling over strong axis?
2. What would be the thickness of a contour encasement protection having a constant thermal conductivity $\lambda_p = 0.12 \, \text{W/mK}$, for a required fire resistance time of 60 minutes?

5.8.2.1   *Fire resistance time of the column with unprotected cross-section*

Classification of the section, see Table 5.2

$$\varepsilon = 0.85 \sqrt{\frac{235}{f_y}} = 0.85$$

Flanges: $c/t_f = 76/9 = 8.44 < 10\varepsilon = 8.5 =>$ The flanges are of Class 2.
Web: $c/t_w = 139/5 = 27.8 < 33\varepsilon = 28.1 =>$ The web is of Class 1.
=> The column cross-section is of Class 2.

The failure of the column occurs at time $t_{fi,d}$ for which the design resistance of the column is equal to the design axial force, for fire situation.

Slenderness at room temperature:

$$\lambda = l_f / i_z = 290 \, / \, 7.23 = 40.11$$

Eulerian slenderness:

$$\lambda_E = \pi \sqrt{E/f_y} = \pi \sqrt{2.1 \times 10^5 / 235} = 93.91$$

Non-dimensional slenderness at room temperature:

$$\bar{\lambda} = \frac{\lambda}{\lambda_E} = \frac{40.11}{93.91} = 0.427$$

The non-dimensional slenderness at elevated temperature is a function of the temperature at failure, which is the unknown in this problem.

In a first iteration it will be approximated by:

$$\bar{\lambda}_\theta = 1.2\bar{\lambda} = 1.2 \times 0.427 = 0.512$$

Imperfection factor:

$$\alpha = 0.65\sqrt{235/f_y} = 0.65\sqrt{235/235} = 0.65$$

$$\varphi_\theta = 0.5(1 + \alpha\bar{\lambda}_\theta + \bar{\lambda}_\theta^2) = 0.5(1 + 0.65 \times 0.512 + 0.512^2) = 0.797$$

$$\chi_{fi} = \frac{1}{\varphi_\theta + \sqrt{\varphi_\theta^2 - \bar{\lambda}_\theta^2}} = \frac{1}{0.797 + \sqrt{0.797^2 - 0.512^2}} = 0.710$$

Design equation:

$$N_{b,fi,t,Rd} = N_{fi,Ed}$$

or $\chi_{fi}k_{y,\theta}Af_y/\gamma_{M,fi} = 410\,\text{kN}$

$$=> k_{y,\theta} = 410\,000/(0.710 \times 3705 \times 235) = 0.663$$

Interpolation in Table II.2 (see Annex II) yields a critical temperature of 528°C for this value of the reduction factor of the effective yield strength.

The same table gives a reduction factor for the Young's modulus $k_{E,\theta} = 0.490$ for this temperature.

Second iteration

$$\bar{\lambda}_\theta = \bar{\lambda}\sqrt{k_{y,\theta}/k_{E,\theta}} = 0.427\sqrt{0.663/0.490} = 0.497$$

$$\varphi_\theta = 0.5(1 + 0.65 \times 0.497 + 0.497^2) = 0.785$$

$$\chi_{fi} = \frac{1}{0.785 + \sqrt{0.785^2 - 0.497^2}} = 0.718$$

$$=> k_{y,\theta} = 410\,000/(0.718 \times 3705 \times 235) = 0.656$$

Critical temperature $\theta_a = 540°C$, see Table II.2

$k_{E,\theta} = 0.484$, see Table II.2

Third iteration

$$\bar{\lambda}_\theta = \bar{\lambda}\sqrt{k_{y,\theta}/k_{E,\theta}} = 0.427\sqrt{0.656/0.484} = 0.497$$

=> The iteration process has converged in two iterations and the critical temperature of 540°C has been obtained.

For I sections under nominal fire, and considering the shadow effect:

$$A_m^*/V = k_{sh}A_m/V = 0.9A_{mb}/V$$

$$A_m^*/V = 0.9 \times 2(h+b)/A = 0.9 \times 2(0.165+0.165)/37.05 \times 10^{-4} = 160\,\text{m}^{-1}$$

Double interpolation in Table I.1 (see Section I) yields:

For $A_m^*/V = 100\,\text{m}^{-1}$, 540°C are obtained after 14.17 min.

For $A_m^*/V = 200\,\text{m}^{-1}$, 540°C are obtained after 9.77 min.

=> For $A_m^*/V = 160\,\text{m}^{-1}$, 540°C are obtained after 11.53 min.

5.8.2.2 Column protected with contour encasement of uniform thickness

The steel temperature corresponding to the failure of the column is $\theta_d = 540$°C. From Table I.2 for protected sections, at $t_{fi,req} = 60$ min., this temperature is obtained for a protection section with $k_p = 1323\,\text{W/m}^3\text{K}$.

For protection with contour encasement:

$$[A_p/V] = \text{steel perimeter cross-section area} = [2b + 2h + 2(b - t_w)]/A$$

$$= [2 \times 0.165 + 2 \times 0.165 + 2(0.165 - 0.005)]/(37.05 \times 10^{-4})$$

$$= 265\,\text{m}^{-1}$$

$$k_p = [A_p/V][\lambda_p/d_p] = 1323\,\text{W/m}^3\text{K}$$

$$=> d_{p,min} = 265 \times 0.12/1323 = 0.024\,\text{m}\,(24\,\text{mm})$$

5.8.3 Fixed-fixed beam supporting a concrete slab

A HE160A beam of 4 m length and with fixed end rotations is supporting a concrete slab. The beam is fabricated from steel of $f_y = 355\,\text{N/mm}^2$ and has design load in fire situation of 9500 N/m. The requested fire resistance for the beam is 30 minutes.

1. Perform the verification in the load domain, in the time domain and in the temperature domain.
2. What would be the failure time, for the steel beam protected by a hollow encasement of 12 mm thickness plates with $\lambda_p = 0.15\,\text{W/mK}$?

Properties of section HE160A: $h = 152\,mm$, $b = 160\,mm$, $t_f = 9\,mm$, $t_w = 6\,mm$,

$$r = 15\,mm, A = 38.77\,cm^2, W_{pl} = 245.1\,cm^3$$

5.8.3.1 Classification of the section, see Table 5.2

$$\varepsilon = 0.85\sqrt{\frac{235}{f_y}} = 0.85\sqrt{\frac{235}{355}} = 0.692$$

Flanges: $c/t_f = (b/2 - t_w/2 - r)/t_f = (160/2 - 6/2 - 15)/9 = 6.89 < 10\varepsilon = 6.92$
The flanges are of Class 2

Web: $c/t_w = (h - 2t_f - 2r)/t_w = (152 - 2 \times 9 - 2 \times 15)/6 = 17.33 < 72\varepsilon = 49.8$
The web is of Class 1:

$\Longrightarrow$ The beam cross-section is of Class 2.

The method for designing the beam is thus to limit the elastically determined bending moment to the plastic bending moment in the sections, see Section 5.6.4.2.

Note: For a Class 1 section, a plastic design of the beam would be performed and the basic design equation in bending would be $M_{fi,t,Rd}^{support} + M_{fi,t,Rd}^{span} = p_{fi,d} L^2/8$

5.8.3.2 Verification in the load domain

For I sections under nominal fire, exposed on three sides and considering the shadow effect:

$$A_m^*/V = k_{sh}A_m/V = 0.9A_{mb}/V = 0.9(2h + b)/A$$

$$= 0.9(2 \times 0.152 + 0.16)/38.77 \times 10^{-4} = 108 \text{ m}^{-1}$$

For $A_m^*/V = 108$ m^{-1} and $t_{fi,req} = 30$ min., Table I.1 gives $\theta_a = 772°C$

For this temperature, Table II.2 gives a reduction factor for effective yield strength $k_{y,\theta} = 0.144$

At the supports

Verification of the design shear resistance

$V_{fi,d} = 9500 \times 4/2 = 19\,000$ N

$(A_m^*/V)_{web} = 2/t_w = 2/0.006 = 333$ m^{-1}

For $t_{fi,req} = 30$ min. and $A_m^*/V = 333$ m^{-1}, Table I.1 gives $\theta_{web} = 834°C$

$=> k_{y,\theta,web} = 0.093$, see Table II.2.

$V_{Rd} = A_v f_y/(\sqrt{3}\ \gamma_{M0}) = 1321 \times 355/(\sqrt{3} \times 1.00) = 270\,750$ N
with $A_v = A - 2bt_f + (t_w + 2r)t_f$

$$= 3877 - 2 \times 16 \times 0.9 + (0.6 + 2 \times 1.5)0.9 = 1321 \text{ mm}^2$$

$$V_{fi,t,Rd} = k_{y,\theta,web} V_{Rd}[\gamma_{M0}/\gamma_{M,fi}]$$

$$= 0.093 \times 270\,750[1.00/1.00] = 25\,180\text{N} > V_{fi,d} = 19\,000 \text{ N}$$

The safety margin is 33%.
   A reduction of $M_{fi,t,Rd}$ allowing for effects of shear is necessary if $V_{fi,d} > 0.5 V_{fi,t,Rd}$ which is the case here.

$$=> \rho = \left(\frac{2 V_{fi,d}}{V_{fi,t,Rd}} - 1\right)^2 = \left(\frac{2 \times 19.00}{25.18} - 1\right)^2 = 0.259$$

Verification of the bending resistance

$$M_{fi,Ed} = pL^2/12 = 9.500 \times 4^2/12 = 12.67 \text{ kNm}$$

$$M_{fi,t,Rd} = \frac{k_{y,\theta}f_y\left[W_{pl} - \dfrac{\rho(h_w t_w)^2}{4t_w}\right]}{k_1 k_2}$$

where:

$k_1 = 0.7$ for unprotected beam exposed on three sides, with a concrete slab on side four,

$k_2 = 0.85$ at the supports of the statically indeterminate beam.

$M_{fi,t,Rd} = 0.144 \times 355 \times [245\ 100 - 0.259(104 \times 6)2/4 \times 6]/(0.70 \times 0.85)$
$= 20.70\ \text{kNm} > 12.67\ \text{kNm}$

=> The design moment resistance of the beam at supports, at time $t_{fi,req} = 30$ min., is higher than the design moment, for fire situation. The safety margin is 63%.

At mid span

$M_{fi,Ed} = pL^2/24 = 9.500 \times 4^2/24 = 6.33\ \text{kNm}$
$M_{fi,t,Rd} = k_{y,\theta}\, f_y\, W_{pl}\, / \,(k_1 k_2)$
$M_{fi,t,Rd} = 0.144 \times 355 \times 245\ 100/(0.70 \times 1.00) = 17.90\ \text{kNm} > 6.33\ \text{kNm}$

where:

$k_1 = 0.7$ for unprotected beam exposed on three sides, with a concrete slab on side four,

$k_2 = 1.00$ if not at the supports.

=> The design moment resistance at mid span, at time $t_{fi,req} = 30$ min., is higher than the design moment, for fire situation. The safety margin is 183%.

5.8.3.3 Verification in the time domain

The critical section is obviously at the supports. It is yet not possible to derive directly the failure time at this section because of the interaction between shear and bending. The situation is also complicated by the fact that different massivity factors have to be considered for the section on one hand and for the web on the other hand. Different possibilities allow determining the fire resistance time.

1. Verification in the load domain can be performed iteratively at various times, until exact equivalence is found between the applied load and the load bearing capacity. In this case, for example, one could check the resistance after 40 minutes of fire and see how the safety margins change for shear and bending at the supports and bending at mid span. Linear interpolation or extrapolation between the values obtained at 30 and 40 minutes should yield the time for which all safety margins are greater or equal to 1.00. If needed, a subsequent verification in the load domain for this new time should confirm the result or give a third point for interpolation. Utilization of computer tools, such as spreadsheet software for example, is highly recommended.

2. It is possible to iterate between the bending and shear resistance at the support. The resistance time could be calculated from the bending resistance at the support, considering the same reduction from shear effects as the one considered at 30 minutes. The shear resistance would then be checked for this new time and the new reduction of the bending resistance would then be evaluated, allowing a new fire resistance time in bending to be determined. The process would be repeated

until convergence. It has to be noted that, because of the second order power of the shear resistance that is taken into account in the reduction of bending resistance, this procedure is not guaranteed to converge.

5.8.3.4 Verification in the temperature domain

Interaction between shear and bending also makes it impossible to determine the critical temperature directly from the utilization factor. Iteration procedures similar to those mentioned above should be applied.

5.8.3.5 Beam protected with hollow encasement

For hollow encasement protection on three sides, with a concrete floor on the fourth side, the section factor is:

$$A_p/V = (2h + b)/A = (2 \times 0.152 + 0.160)/(38.77 \times 10^{-4}) = 120 \, \text{m}^{-1}$$
$$[A_p/V] \, [\lambda_p/d_p] = 120 \times 0.15/0.012 = 1\,500 \, \text{W/m}^3\text{K}$$

For this section protected by a hollow encasement, it is assumed that the temperature of the web is equal to the uniform temperature in the section.

Iteration 1
Let us assume that $k_{y,\theta} = 0.088$

$$V_{fi,t,Rd} = k_{y,\theta,web} V_{Rd}[\gamma_{M0}/\gamma_{M,fi}] = 0.088 \times 270\,750 = 23.83 \, \text{kN} > V_{fi,d} = 19 \, \text{kN}$$

$$\Rightarrow \rho = \left(\frac{2 \times 19.00}{23.83} - 1\right)^2 = 0.354$$

Bending at the support:

$$M_{fi,t,Rd} = 0.088 \times 355 \times [245\,100 - 0.354(104 \times 6)^2/4 \times 6]/(0.70 \times 0.85)$$
$$= 12.57 \, \text{kNm} < 12.67 \, \text{kNm}$$

Iteration 2

$$k_{y,\theta} = 0.089$$
$$V_{fi,t,Rd} = 0.089 \times 270\,750 = 24.10 \, \text{kN} > V_{fi,d} = 19 \, \text{kN}$$

$$\Rightarrow \rho = \left(\frac{2 \times 19.00}{24.10} - 1\right)^2 = 0.333$$

Bending at the support:

$$M_{fi,t,Rd} = 0.089 \times 355 \times [245\,100 - 0.333(104 \times 6)^2/4 \times 6]/(0.70 \times 0.85)$$
$$= 12.73 \, \text{kNm} > 12.67 \, \text{kNm}$$
The value of 0.089 for $k_{y,\theta}$ satisfy the load bearing requirement.

For this value, Table II.2 gives a failure temperature of 842°C.

Interpolations in Table I.3:

For $k_p = 1200\ \text{W/m}^3\text{K}$, $842°\text{C}$ is obtained after $164\,\text{min}$.
For $k_p = 2000\ \text{W/m}^3\text{K}$, $842°\text{C}$ is obtained after $116\,\text{min}$.
$=>$ For $k_p = 1500\ \text{W/m}^3\text{K}$, $842°\text{C}$ is obtained after $146\,\text{min}$.

5.8.4 Class 3 beam in lateral torsional buckling

A simply supported beam of span of 6 m and of section HE180A, $f_y = 355\,\text{N/mm}^2$, is supporting a secondary beam at mid span. The secondary beam induces in the main beam a concentrated load of 20 kN in fire situation.

What is the temperature at failure of the main beam?

Properties of section HEA180: $h = 171\,\text{mm}$, $b = 180\,\text{mm}$, $t_f = 9.5\,\text{mm}$, $t_w = 6\,\text{mm}$, $r = 15\,\text{mm}$, $W_{el,y} = 293.6\,\text{cm}^3$, $I_z = 924.6\,\text{cm}^4$, $I_w = 60\,210\,\text{cm}^6$, $I_t = 14.8\,\text{cm}^4$

Classification of the section, see Table 5.2

$$\varepsilon = 0.85 \sqrt{\frac{235}{f_y}} = 0.85 \sqrt{\frac{235}{355}} = 0.692$$

Flanges: $c/t_f = (b/2 - t_w/2 - r)/t_f = (180/2 - 6/2 - 15)/9.5 = 7.57 < 14\varepsilon = 9.69$
The flanges are of Class 3.

Web: $c/t_w = (h - 2t_f - 2r)/t_w = (171 - 2 \times 9.5 - 2 \times 15)/6 = 20.33 < 72\varepsilon = 49.8$
The web is of Class 1.

$\Longrightarrow$ The beam cross-section is of Class 3.

$M_{fi,Ed} = q_{d,fi} L/4 = 20 \times 6/4 = 30\,\text{kNm}$

Failure is reached when:

$M_{b,fi,t,Rd} = \chi_{LT,fi}\, W_{el,y} k_{y,\theta,com} f_y / \gamma_{M,fi} = M_{fi,Ed} = 30\,\text{kNm}$

From which:

$k_{y,\theta,com} = M_{fi,Ed} \gamma_{M,fi} / (\chi_{LT,fi} W_{el,y} f_y)$

The critical elastic moment of the beam will be needed for checking stability with regard to lateral torsional buckling. For a simply supported beam with I section, the critical elastic moment is computed according to Annex F of ENV 1993-1-1.

$$M_{cr} = C_1 \frac{\pi^2 E I_z}{L^2} \left[\sqrt{\frac{I_w}{I_z} + \frac{L^2 G I_t}{\pi^2 E I_z} + (C_2 y_g)^2} - (C_2 y_g)^2 \right] = 210.9\ \text{kNm}$$

with:

$L = 6/2 = 3\,\text{m}$, considering that the secondary beam provides lateral support at mid span,
$y_g = 0.5\,h = 85.5\,\text{mm}$
$C_1 = 1.365$, $C_2 = 0.553$ (for concentrated load in the middle span).

Non dimensional slenderness at room temperature

$$\bar{\lambda}_{LT} = \sqrt{\frac{W_{el,y}f_y}{M_{cr}}} = \sqrt{\frac{293.6 \times 355}{220\,900}} = 0.703$$

Iteration 1
In the first iteration, the non-dimensional slenderness at elevated temperature will be estimated by:

$$\bar{\lambda}_{LT,\theta,com} = 1.2\,\bar{\lambda}_{LT} = 1.2 \times 0.703 = 0.844$$

$$\alpha = 0.65\sqrt{235/f_y} = 0.529$$

$$\Phi_{LT,\theta,com} = 0.5[1 + \alpha\bar{\lambda}_{LT,\theta,com} + (\bar{\lambda}_{LT,\theta,com})^2]$$
$$= 0.5[1 + 0.529 \times 0.844 + 0.844^2] = 1.079$$

$$\chi_{LT,fi} = \frac{1}{\Phi_{LT,\theta,com} + \sqrt{[\Phi_{LT,\theta,com}]^2 - [\bar{\lambda}_{LT,\theta,com}]^2}}$$

$$= \frac{1}{1.079 + \sqrt{1.079^2 - 0.844^2}}$$
$$= 0.571$$

$$\Rightarrow k_{y,\theta,com} = M_{fi,Ed}\gamma_{M,fi}/(\chi_{LT,fi}W_{el,y}f_y)$$
$$= 30 \times 10^6/(0.571 \times 293\,600 \times 355) = 0.504$$

with the corresponding temperature $\theta_{a,com} = 589°C$, see table II.2.

Iteration 2
In the second iteration, $\chi_{LT,fi}$ is computed considering the temperature determined in the first iteration.

For $\theta_{a,com} = 589°C$, $k_{E,\theta,com} = 0.340$, see Table II.2

$$\bar{\lambda}_{LT,\theta,com} = \bar{\lambda}_{LT}\sqrt{k_{y,\theta,com}/k_{E,\theta,com}} = 0.703\sqrt{0.504/0.340} = 0.856$$

$$\Phi_{LT,\theta,com} = 0.5[1 + 0.529 \times 0.856 + 0.856^2] = 1.093$$

$$\chi_{LT,fi} = \frac{1}{1.093 + \sqrt{1.093^2 - 0.856^2}} = 0.564$$

$$\Rightarrow k_{y,\theta,com} = 30 \times 10^6/(0.564 \times 293\,600 \times 355) = 0.510$$

with the corresponding temperature $\theta_{a,com} = 587°C$, see table II.2.
Because the difference between the temperatures calculated in the two iterations is less than 0.4%, the iteration process may stop. The temperature of the compression flange, corresponding to the failure of the beam is $\theta_{a,com} = 587°C$.

Chapter 6

Joints

6.1 General

Joints, also sometimes referred to as connections, play an important role in a structural framing since they transfer forces from one member to another. The performance of connections is highly important under extreme loading conditions since the critical regions of a steel structure are in plastic state and also due to the greater need for redistribution of forces from one critical region to another critical region. In most fire situations, it is likely that some parts of the structure in the non-fire exposed regions have much greater capacity than the fire exposed regions and thus the fire resistance of the structure is highly dependent on the extent of redistribution (through connections) that occurs from highly stressed regions to less stressed regions. Thus the performance of connections is crucial for the stability of structural systems in buildings when they are exposed to fire. This aspect, of the importance of connections, has been observed and documented (FEMA 2002) in the damaged WTC buildings around Ground Zero as a result of fires initiated after the collapse of twin-towers.

In modern steel-framed buildings, connections between various members may be either of bolted or welded construction, or a combination of these types. Most codes and standards require steel connections to be provided with some level of fire protection. However, many codes do not explicitly state fire resistance requirements for connections. Further, in establishing fire ratings through prescriptive approaches, connections are generally not included as part of the assembly tested in traditional fire-resistance tests. Furthermore, most modelling efforts assume that the pre-fire characteristics of a connection are preserved during the fire exposure.

US codes generally give little guidance on the fire design of connections. No explicit provisions are specified in US codes and standards. A closer look at the overall fire resistance provisions clearly imply that the connections should be protected to the same level of fire resistance as that of the connecting members.

In Europe, it was also commonly assumed since the 70's that there is no need to take special provisions for the connections as long as they are protected at least in the same manner as the adjacent members that they connect (ECCS, 1983). This implies that, if none of the connected members is protected, then there is no need to protect the joint. This concept was based on the idea that the thermal massivity of the joint should be higher than the massivity of the members because of the presence of additional mass in the connection zone, either from end plates, fin plates, web cleats or stiffeners. It was also based on the observation of numerous unprotected steel structures that completely collapsed in severe fire, and where the steel beams were severely distorted, but rarely detached from the columns. The perception has evolved during the last decade because of the appearance of the conceptual design of multi-storey buildings where the columns are protected while the beams and the joints are not. The demand on the connections in such systems is of course much higher. This is especially the case when axial restraint induces axial forces in the beams and thus in the connections. Compression forces first develop in the beams due to restraint against thermal expansion and tension forces may develop in a later stage when significant vertical deflections in the beams transform the beams from elements in bending to elements in tension, more like cables. More tension can even develop in the beams, and thus in the joints, in the cooling phase of a natural fire. It has also been observed that not only the resistance to the varying forces, but also the ductility of connections must be very important in order to accommodate the large rotations linked to the large displacements that develop when the beams act in a catenary mode. A good review of recent research performed on the behaviour of joints in the fire situation may be found in Al-Jabri et al (2008).

However, in Eurocode 3, design rules for both bolted and welded connections in the fire situation are only specified through the introduction of strength reduction factors. Moreover, simplified temperature distribution of joints in the fire situation is also proposed, which may be used in strength analysis. This approach, though only primitive in nature, is rational and one step ahead of the "connection provisions" in other Codes of practice. This Chapter is devoted to highlighting the Eurocode methodology for the design of connections.

6.2 Simplified procedure

Eurocode 3 states in the main text that the fire resistance of a bolted or a welded joint may be assumed to be sufficient, provided that the three following conditions are satisfied:

- The joint has at least the same fire protection as all connected members. In particular, this means that it is not necessary to verify the joints in an unprotected steel structure if the other conditions, especially the one regarding the utilisation of the joint, are fulfilled.
- The utilisation of the joint is equal or less than the highest value of utilisation of any of the connected members.
- The resistance of the joint at room temperature satisfies the recommendations given in EN 1993-1.8. This condition that the design must comply with the rules at ambient temperature is not specific to joints, as it is applicable for every part of the steel structure.

The justification usually given for this recommendation is that due to the additional material in the joints and also to the shadow effects created by the connected members, the temperatures are lower in the joints than those within the adjacent members. Another explanation for the lower temperatures in joints is linked to the geometry of the fire compartment, that results in lower temperatures at the corners, where the joints are usually located. This idea of lower temperatures at the joints is a similar concept to the one leading to the utilisation of the reduction factor k_2 that is considered at the supports of statically indeterminate beams, see Section 5.5.4.2.2. But if this is generally true for the beam-column connections of frame structures, it is certainly not the case for a continuity joint in the middle of the lower chord of a roof truss, for example.

For the same reason, the Eurocode states that when verifying connection resistance, the net-section failure at fastener holes need not be considered, provided that there is a fastener in each hole. However, simple numerical simulations yet show that this provision of not considering net section failure is not valid in general (Franssen and Brauwers, 2002). In fact, if the standard temperature-time curve is used, there is a tendency for the temperatures in steel to level off after a certain amount of time and it requires a very significant difference in thermal massivity in order to produce a noticeable temperature difference, see Figure I.4. This is especially true in the case of long fire durations and for thermally protected members, where the benefit yielded by a somewhat lower temperature in the section with the fastener may well be totally overwhelmed by the reduction of net section. Therefore, designer are well advised to perform this verification when the resistance of this net-section is critical for the stability of the structure.

Eurocode does not define in the section on design of joints the utilisation factor for the application of the second condition. It has to be assumed that the same definition has to be used as the one mentioned in the section dealing with the design in the temperature domain, see Equation 5.42, given here again for clarity reasons.

$$\mu_0 = E_{fi,d}/R_{fi,d,0} \tag{5.42}$$

This equation implies that, in order to be allowed not to calculate the fire resistance of the joint, the fire resistance of the joint at time $t = 0$ has to be calculated first! It is then no miracle that, as a further simplification of the method, it is stated that the comparison of the level of utilisation within the joints and joined members may be performed for room temperature. This means that Equation 6.1 will be used for the utilisation of the joints and of the adjacent members.

$$\mu = E_d/R_d \tag{6.1}$$

Due care must be taken when applying this method to one particular situation. Let us assume that the starting point is a design at room temperature, in which the joints and the connected elements were normally proportioned. If all connected elements satisfy the fire resistance requirement, the recommendation above implies that the joints also meet this requirement. If, on the contrary, the fire resistance time of one element, a beam for example, is not sufficient and instead of applying a fire protection it is preferred to use a stronger section for this element, it is not to be expected that the resistance of the joint will increase in the same proportion. Consequently, if the elements are overdesigned for fire resistance purposes, the adjacent joints must be overdesigned accordingly. The new value of utilisation for the overdesigned elements must be the reference when the resistance of the joint is revisited.

As an alternative to the above mentioned method, the fire resistance of a joint may be determined using the method given in Annex D of Eurocode 3. This method is presented and discussed in the following section.

6.3 Detailed analysis

When assessing the fire resistance of a joint, the temperature distribution of the joint components should be evaluated and then the resistance of each component at high temperature should be determined accordingly.

6.3.1 Temperature of joints in fire

As an alternative to the above simplified method, the informative Annex D in Eurocode states that, in order to compute the fire resistance of the components, the temperature of a joint may be assessed using the local thermal massivity of the parts forming that joint as shown, for example, in ECCS (2001). In a more simplified way, a uniform distributed temperature may be assessed within the joint; this temperature may be calculated using the minimum value of the thermal massivity of the steel members connected to the joint.

A more refined procedure is given for beam to column and beam-to-beam joints, where the beams are supporting any type of concrete floor. In this case, the temperature distribution in the joint may be obtained from the temperature of the bottom flange at mid span. The following formulas are recommended for computing the temperature of joint components.

$$\theta_h = 0.88\,\theta_0[1 - 0.3h/D] \qquad\qquad \text{if } D \le 400\,\text{mm} \qquad (6.2)$$

$$\theta_h = 0.88\,\theta_0 \qquad\qquad \text{for } h \le D/2$$
$$\qquad\qquad\qquad\qquad\qquad\qquad\qquad\qquad \text{if } D > 400\,\text{mm} \qquad (6.3)$$
$$\theta_h = 0.88\,\theta_0[1 + 0.2(1 - 2h/D)] \quad \text{for } h > D/2$$

where:
θ_h is the temperature at height h in the steel beam,
θ_0 is the bottom flange temperature of the beam remote from the joint,
    i.e. based on the thermal massivity of the section heated on 4 sides, see Table 5.3.
h is the height of the component being considered above the bottom of
    the beam,
D is the depth of the steel section.

6.3.2 Design resistance of bolts and welds in fire

Verification is based on the strength as determined at room temperature multiplied by the reduction factors for strength of bolts and welds given in Table 6.1. These factors are based on the research by Kirby (1995) and Latham and Kirby (1990).

6.3.2.1 Bolted joints in shear

The shear design resistance of an individual bolt in fire should be determined from Equation 6.4:

$$F_{v,t,Rd} = F_{v,Rd}k_{b,\theta}\frac{\gamma_{M2}}{\gamma_{M,fi}} \qquad\qquad\qquad (6.4)$$

Table 6.1 Strength reduction factors for bolts and welds at various temperatures

θ_a [°C]	$k_{b,\theta}$	$k_{w,\theta}$
20	1,000	1,000
100	0,968	1,000
150	0,952	1,000
200	0,935	1,000
300	0,903	1,000
400	0,775	0,876
500	0,550	0,627
600	0,220	0,378
700	0,100	0,130
800	0,067	0,074
900	0,033	0,018
1000	0,000	0,000

where:

$F_{v,Rd}$ is the design shear resistance of the bolt per shear plane calculated, at normal temperature, assuming that the shear plane passes through the threads of the bolt (from Table 3.4 of EN 1993-1-8),

$k_{b,\theta}$ is the reduction factor determined for the appropriate bolt temperature given in Table 6.1,

γ_{M2} is the partial safety factor at normal temperature,

$\gamma_{M,fi}$ is the partial safety factor for fire conditions.

In the computation of $F_{v,Rd}$, it must be noted that, irrespective of whether the shear plane passes through the threaded or unthreaded portion of the bolt, the shear area of the bolt in case of fire must be considered as the tensile stress area of the bolt A_s, i.e. assuming that the shear plane passes through the threads of the bolt.

The design bearing resistance of an individual bolt in fire should be determined from Equation 6.5:

$$F_{b,t,Rd} = F_{b,Rd} k_{b,\theta} \frac{\gamma_{M2}}{\gamma_{M,fi}}$$ (6.5)

where $F_{b,Rd}$ is the design bearing resistance at room temperature, determined from Table 3.4 of EN 1993-1-8.

The slip resistant type joints are not efficient in case of fire and therefore should be considered as having slipped in fire. Consequently, the resistance of a single bolt should be verified as for a usual bolt in shear with the formulas above.

6.3.2.2 Bolted joints in tension

The design tension resistance of an individual bolt in fire should be determined from Equation 6.6:

$$F_{ten,t,Rd} = F_{t,Rd} k_{b,\theta} \frac{\gamma_{M2}}{\gamma_{M,fi}}$$ (6.6)

where $F_{t,Rd}$ is the design tension resistance of the bolt at room temperature, determined from Table 3.4 of EN 1993-1-8.

6.3.2.3   Fillet welds

The design resistance per unit length of a fillet weld in fire should be determined from Equation 6.7:

$$F_{w,t,Rd} = F_{w,Rd} k_{w,\theta} \frac{\gamma_{M2}}{\gamma_{M,fi}} \tag{6.7}$$

where:
$k_{w,\theta}$ is obtained from Table 6.1 for the appropriate weld temperature,
$F_{w,Rd}$ is the design strength of the fillet weld at normal temperature, determined from Clause 4.5.3 of EN 1993-1-8.

6.3.2.4   Butt welds

For normal temperatures, the design strength of a full penetration butt weld should be taken as equal to the design resistance of the weakest connected parts, provided that the conditions in 4.7.1.(1)/EN1993-1-8 are fulfilled.

The fire design strength of a full penetration butt weld is computed using the design strength at normal temperature, corrected by the following reduction factors:

- For temperatures up to 700°C, the reduction factors for structural steel in fire situation.
- For temperatures above 700°C, the reduction factors $k_{w,\theta}$ given in Table 6.1.

Chapter 7

Advanced Calculation Models

7.1 General

Much of the discussion presented in previous chapters is focused on application of simple calculation approaches for tracing the fire response of structural members. These simple calculation techniques, though one step above the prescriptive based approaches, may not provide realistic fire response of structural system since a number of factors cannot be accounted for in simplified analysis. In lieu of these simple calculation methods, advanced calculation models can come handy for predicting the realistic fire response of structural response. These advanced calculation models are generally based on finite element techniques and can account for geometric and material nonlinearity, in addition to various high temperature effects, such as creep and transient strains, and restraint effects. The accuracy of the predictions is often dependent on the level of the complexity of the model adopted for the analysis, discretization utilized in the analysis, and on the accuracy of the input (material) data. In addition, the application of these advanced calculation models requires significant expertise, time and computational effort.

As discussed in Chapter 2, advanced calculated models could be applied to a single member, an assembly, or the entire building frame. The US codes do not specifically provide requirements or guidance on the use advanced calculated models. No specific guidelines are provided in North American codes such as IBC and NBCC, on the analysis procedure, or high temperature properties material models or design fires. However, the recent edition of AISC design manual (AISC 2005) describes the requirement for advanced methods of structural analysis for performance-based fire resistant design. It also recommends the use of Eurocode material properties for carrying out advanced analysis. Thus it is possible in North America to get regulatory approvals by undertake fire resistance analysis using the advanced calculation models.

Eurocodes are more pro-active in facilitating the use of advanced calculation models for fire resistance design and analysis. These codes provide detailed procedure to be used for analysis, constitutive models for high temperature material properties, as well as various parametric (design) fire curves. The details of these provisions are discussed in this Chapter.

7.2 Introduction

Advanced calculation models are defined in Eurocode 3 as *"design methods in which engineering principles are applied in a realistic manner to specific applications"*. This definition is quite vague, to say the least.

Eurocode provide additional information in Section 4.3 that deals specifically with advanced calculation models. For example, it is stated that *"advanced models should be based on fundamental physical behaviour"*. This means that, for example, equations derived from best fit with experimental results cannot be labelled as advanced models.

The Eurocode states that advanced calculation models *should* include separate models for the determination of temperature in the structure, on one hand, and the mechanical behaviour of the structure, on the other hand. This is indeed the procedure adopted in most calculation advanced models nowadays; most models are made of two separate sub-models. It fact, there is no reason why a model that has as sole objective the determination of temperatures in structures subjected to fire would not deserve the title of advanced model. The same could be said of a mechanical model that has to rely on other means for the determination of the temperature distribution. On the other hand, there is no reason why a model that would perform a fully coupled thermo-mechanical analysis could not deserve the title of advanced model. For some problems such as, for example, spalling in concrete, the integrated models are the only possible hope to achieve any result. To the knowledge of the authors, such coupled models have not been used for steel structures so far, but developers should not be forbidden to try. Too much importance must probably not be paid to the restrictive character of this word "should" that is present in the Eurocode.

Eurocode states that advanced models *may* be used with any heating curve, provided that the material properties are known for the relevant temperature range. In fact, the sentence should be written or, at least, understood the other way around: advanced models should not be used with heating curves for which the material properties are not known for the relevant temperature range. The heating rate of steel that must be comprised between 2 and 50 K/min for the Eurocode material model to be valid is one point on which the Eurocode draws the attention. There is yet another important point on which the Eurocode gives no guideline for the use of advanced models, namely the cooling phase that is present in parametric or zone fire models. No recommendation is given about the mechanical properties of steel during the cooling phase. Therefore designers have to make their own judgement and decision about the reversible character of, for example, the yield strength of steel or its thermal elongation.

Advanced calculation methods may be used with any type of cross-section. Here again, there is no reason why a model dedicated to a particular type of sections should not deserve the name of advanced model, provided that all other requirements related to fundamental physical behaviour be fulfilled. The only factor is that such a model would have a field of application limited to this type of section.

Practically speaking, application of the advanced calculation model requires utilization of sophisticated non-linear numerical computer software. The finite element method appears to be the method of choice for thermal as well as for mechanical analyses but finite differences have also been used in thermal analyses. Finite volumes and boundary elements could also be considered, although they have rarely been used in structural fire engineering applications until now.

More information about advanced calculation models is given in the subsections of the Eurocode dedicated first to the thermal analysis and, then, to the mechanical analysis. These provisions are mentioned and discussed in the following sections.

7.3 Thermal analysis

7.3.1 General features

Eurocode states that advanced models for the thermal response should be based on the acknowledged principles and assumptions of the theory of heat transfer. This sentence is nothing more than a particularization of the concept of "*fundamental physical behaviour*" to the case of temperature calculations.

Advanced models should consider the variation of the thermal properties of the material with temperature. The fact that the term "*material*" is singular, plus the comment at the end of the sentence "*see section 3*", i.e. "*see the section on material properties of steel*", seem to indicate that only thermal properties of steel must be taken as temperature dependent. Thermal properties of insulating materials, on the contrary, may then be taken as temperature dependent or not, depending of the outcome of experimental test results made to determine these properties.

Eurocode specifies that the effects of non-uniform thermal exposure and of heat transfer to adjacent building components may be included where appropriate. Non-uniform thermal exposure is indeed an essential feature of the design if any localized fire model is used. Heat transfer to adjacent building may be considered if the cooling effect that would result is expected to have a significant influence on the mechanical behaviour, see Section 7.3.4.

According to Eurocode provisions, the influence of moisture content and moisture migration within the fire protection material may conservatively be neglected. In fact, in most advanced models practically used, moisture is taken into account implicitly by a suitable modification of the apparent specific heat. Some models take into account the energy of evaporation explicitly. Moisture movements are rarely modelled. It has to be understood that, if any moisture movement in the insulation may have a slight influence on the temperature elevation of steel around and below 100°C, these effects are completely diluted and damped when steel reaches the temperature levels usually associated to failure, typically several hundreds of degrees centigrade.

The most important point, for a designer using such an advanced calculation model, is to understand the capabilities offered by the model, as well as the limitations inherent to the model. They can differ from one model to the other. The following sections nevertheless present some aspects that are thought to be rather general.

7.3.2 Capabilities of the advanced thermal models

The main advantage of a thermal advanced model is its ability to determine the *non-uniform temperature distribution* in sections or members. There is thus no need to make the hypothesis that the temperature is uniform, to calculate local thermal massivity in different parts of a member, or to wonder where the maximum temperature in the section is. Each point in the member has its own temperature. Figure 7.1 shows the isotherms calculated in a steel hot rolled HE400M section after 10 minutes of ISO fire. There is a difference of some 120°C between the coldest part of the section, namely the

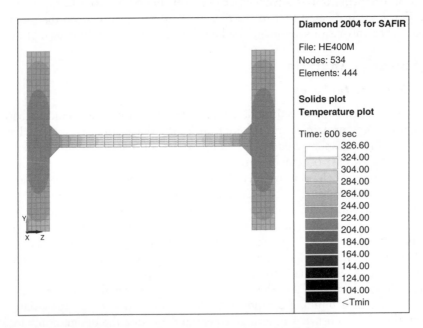

Fig. 7.1 Isotherms in a steel section resulting from thermal analysis

junction between the web and the flanges, and the hottest part of the section, namely the centre of the web.

Heat sink effects from concrete slabs supported by a steel beam are considered in detail. Figure 7.2 shows the isotherms as calculated in a steel section supporting a concrete slab on the upper flange. Note that the scale of grey in the figure has been chosen from 410 to 910°C in order to highlight the differences in the steel section and in the lower part of the slab. This is the reason why the upper part of the slab appears as very black.

A great advantage of the finite element method is the total *versatility* of the geometries that can be considered. Nearly all sections or members encountered in civil engineering applications can be represented by linear triangles or quadrangles, for 2D analyses, or by volume elements with 6 or 8 nodes for 3D analyses. The same software can thus allow considering **H** or **I** sections, rectangular or circular hollow steel sections, angles, or complex 3D joints, with or without thermal protection, either as a contour or as a hollow encasement. The finite difference method is in this respect somewhat more restrictive in the sense that it is more adapted to structured meshes and, therefore, to regular and repeatable shapes.

As mentioned above, *3D analyses* can be performed in order to calculate the temperature distribution in complex assemblies, although at a higher price in term of computing time and, even more, in term of time required for the discretisation. Figure 7.3 shows the discretisation of the structure that has been created in order to analyse the temperature distribution in a beam to column joint with a concrete slab. The column is protected by concrete located between the flanges. Only quarter of the assembly is modelled owing to the presence of two vertical planes of symmetry.

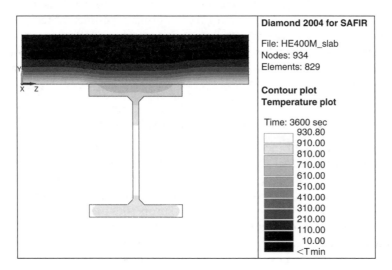

Fig. 7.2 Isotherms in a steel section supporting a concrete slab resulting from thermal analysis

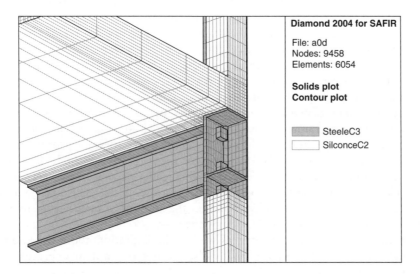

Fig. 7.3 Discretisation of beam-column joint with a concrete slab for thermal analysis

The *quantity of data* provided by such an analysis is enormous. It is as if a thermal sensor was present at every node of the numerical model and there are easily several hundreds of them in a 2D analysis and several thousands of them in a 3D analysis. This allows, with the aide of now commonly available graphic interface software, plotting isotherms on the structure at every time interval during the fire, something that would be unfeasible from the result of a limited number of experimental measurements.

The *quality of the results* is also remarkable. Bearing in mind all hypotheses and limitations mentioned otherwise, it has to be recognised that these virtual thermal sensors normally do not malfunction during the course of the fire, and do not produce results that suddenly diverge and make no sense compared to the measurements of the other sensors, whereas this happens regularly during experimental tests. This is the case of course if the software has been used appropriately and this depends very much on the expertise of the user, but this expertise can be gained or learnt. The location of each node is also perfectly known without any uncertainty, which is not always the case in experiments.

7.3.3   Limitations of the advanced thermal models

Amazingly enough, one of the major limitations that designers are facing in order to perform a non-linear thermal analysis is perhaps the lack of reliable information on the *thermal properties* of insulating materials. Not all companies producing these types of products have dedicated necessary resources required to perform the experimental test programs yielding the high temperature thermal properties used in numerical analyses. Most of them simply rely on graphical design aids that are a summary of the test results, in the form, for example, of required product thickness as a function of the thermal massivity of the section and, sometimes, the critical temperature. It has to be mentioned that this limitation also applies to simple calculation models.

Another important limitation is that, with nearly all software currently available, *the geometry* of the structure in which the temperatures will be determined *is fixed* by the user and it will not vary during the analysis. Spalling of concrete, for example, is normally not predicted nor taken into account by such software. Falling off of gypsum plaster boards from steel studs are also modelled or taken into account with great difficulty. The increase in thickness of intumescent insulation products is also normally not represented explicitly. There are some possibilities to consider these effects in an approximate manner, depending on the expertise of the user and on the capabilities of the software. For example, it is possible to model an intumescent product by a constant thickness layer of a product with "*equivalent*" thermal properties. In the software SAFIR developed at the University of Liege (Franssen 2003), it is possible to restart a thermal analysis at a defined time considering that one or several layers of finite elements have suddenly disappeared from the model. This can be useful in certain cases, but it has to be understood that the amount of material to suppress as well as the time to do it must be decided by the user.

The limitation of a fixed geometry, as well as the limitation about the knowledge of thermal properties of insulating materials, are not relevant in unprotected steel structures.

When different elements are in contact with each other, it is usual to assume a *perfect thermal contact*, which is an approximation. This can be the case between the steel of a hollow section and the concrete that is inside, or between a steel beam and the concrete slab that it may support. This can also be the case between different steel components in a joint. If appropriate information is available about the contact resistance, it is of course possible to model it by a very thin layer of elements with appropriate thermal properties (Renaud 2003), but this is not commonly done.

Boundary conditions are easily taken into account when the fire is represented by a temperature-time curve representing the condition of the gases surrounding the structure, as is the case, for example, with nominal fires, with parametric fires or zone models. It is also possible to take into account a prescribed impinging heat flux as is the case, for example for the localized fire model of Eurocode 1 although, in this case, the variability of the flux in space makes the procedure somehow more complex. The complexity is yet of a higher order of magnitude if such a model for determining the temperature in the structure has to be *interfaced with CFD* software, "*Computational Fluid Dynamics*". Such complex models indeed describe the situation in the fire compartment by an enormous amount of information such as the temperature in virtually every point in the compartment, plus radiation intensities at every point from every direction. However, this information is transferred to the thermal model for determining the temperatures in the structure is still a topic of research at present (Welsch et al., 2008).

7.3.4 Discrepancies with the simple calculation models

It has to be accepted that some discrepancies would exist between the results provided by an advanced and a simple calculation model because of the approximations that have to be introduced in the latter one. Normal philosophy, when introducing approximations or simplifications, is that they are on the conservative side. Yet, if an advanced calculation model is used for calculating the temperature distribution in a concave unprotected steel section and if the nominal boundary conditions prescribed in Eurocode 1 are applied as such on the whole perimeter area of the section, the temperatures found by the advanced model will be higher than the temperature found with the simple model. This goes against the usually accepted principle that a simplified model should be on the conservative side compared to a more advanced model. Although this is not specifically mentioned in the Eurocode, the authors recommend that one of the following procedures be used when using the advanced calculation model in order to introduce the concept of the shadow effect in the model.

1. The simplest procedure is to multiply the value of the coefficient of convection and the emissivity of steel by the value of k_{sh} on the whole perimeter of the section. This will ensure the same boundary conditions for the advanced model as that for the simple model.
2. A more refined procedure is to calculate in each concave part of the section the view factors between, on one hand, each of the surfaces of the real section where the energy arrives and, on the other hand, the surface of the box contour through which this energy passes. The coefficient of convection (α_c) and the emissivity of the material (ε_m) are then multiplied by the relevant view factor depending on their position. Figure 7.4 shows an example of the multiplicative coefficients that could be found in a hypothetical I section.

Each view factor between the internal surface of the flanges and the web, on one hand, and the surface of the box contour around the flanges, on the other hand, is calculated according to the rule of Hottel. This rule, valid in a 2D situation, is given by Equation 7.1 with the symbols defined in Figure 7.5.

$$F_{ij} = \frac{\overline{AD} + \overline{BC} - \overline{AC} - \overline{BD}}{2\,\overline{AB}} \tag{7.1}$$

Except in very re-entrant concave sections, the first procedure should yield results that are very close to those yielded by the most sophisticated second procedure.

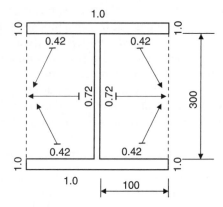

Fig. 7.4 Coefficients for multiplying the boundary conditions in an I section

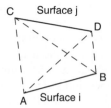

Fig. 7.5 The rule of Hottel

7.4 Mechanical analysis

7.4.1 *General features*

Advanced calculation models for the mechanical response should be based on the acknowledged principles and assumptions of the theory of structural mechanics. This sentence is a particularization of the concept of "*fundamental physical behaviour*" to the case of structural analyses.

The changes in mechanical properties of steel with temperature should be taken into account, obviously. Temperature dependent mechanical properties have to be taken into account.

The effects of thermal expansion should be taken into account. This is done at the level of the constitutive model used in analyses software to represent the behaviour of the material. A component accounting for thermal strain has to be explicitly introduced, which is standard practice.

The combined effects of mechanical actions, thermal actions and geometrical imperfections have to be taken into account. If the structural analysis is performed for the structure under design load in case of fire and the effects of temperatures on the material properties are taken into account, the first two effects are automatically combined. Geometrical imperfections have to be introduced. It is logical to introduce the same imperfections as for ambient temperature design because they represent the initial state of the structure, independent of the load case or of the action, normal action or accidental action. In fact, geometrical imperfections of members are not so critical if the member is submitted to bending moments or to thermal gradients across the depth of the section because these effects will generate transverse displacements that are of an order of magnitude higher than initial imperfections. For example, the Eurocode states that a sinusoidal imperfection of maximum 1/1000 of the length of the bar should be applied when not specified otherwise by the relevant product standard. But this provision is only for a steel isolated vertical member. It is likely that the meaning of "*isolated vertical member*" should be understood as "*simply supported member submitted to axial loading*". If the member is also subjected to bending moments, these will induce first order lateral displacements which are probably of an order of magnitude higher than initial imperfections.

Geometrical non-linear effects have to be taken into account. This can be understood as a requirement that large displacements must be taken into account explicitly. Indeed, these displacements are usually very large in the fire situation. They may generate second order effects that lead to failure, as in a cantilever fire wall heated on one side for example. On the contrary, large displacements may allow the structure to find another load path for supporting the loads, as is the case, for example, in members that are mainly submitted to bending at room temperature and may develop a catenary effect in the fire situation. Taking large displacements implicitly as, for example, by a buckling coefficient in member under compression, is not an advanced method but a simple calculation method.

It is for example well known that, in some conditions, although unprotected steel beams may loose nearly all their load bearing capacity in bending because of the high temperatures that they experience, the concrete floor that they support may be sufficient to withstand the applied loads. This is due to the tensile membrane forces that develop in the steel mesh present in the slab. This effect is not presented in details in this book because, although its existence has been shown by the steel industry, it has more to do with the design of a concrete slab or a composite steel-concrete assembly.

The effects of non-linear material properties have to be taken into account including, states the Eurocode, "*the unfavourable effects of loading and unloading on the structural stiffness*". The material model proposed by the Eurocode for steel is indeed highly non-linear. It is important, in the opinion of the authors, that the full stress-strain relationship be represented, including the descending branch. If not, excessively ductile behaviour could be predicted, linked to very high displacements. The effects of loading and unloading on the structural stiffness mean that the material model is not elastic. For steel, a plasticity model is normally used. On Figure 7.6, for example, if the material has experienced a loading from O to A on the virgin curve for a defined temperature and the strain thereof decreases, the unloading will follow the path A-B and not the reversible path A-O. This schematic example is shown here for a constant temperature. Somewhat more complex algorithmic considerations have to be taken into account and implemented if the temperature as well as the strain and the stress change simultaneously, see Franssen (1990) for increasing temperature, and El-Rimawi et al. (1996), the generalisation for the cooling phase. Whether these effects have favourable or unfavourable effects is not certain. Anyway, they have to be implemented in the constitutive model.

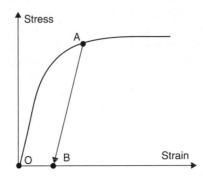

Fig. 7.6 Material plasticity model illustrating loading and unloading phases

The previous requirement on unloading has as a consequence the fact that a step by step or a transient analysis of the structure should be performed, with the time progressively increasing. In fact, because of thermal effects and large displacements, strain reversals and unloading in the material occur continuously in a structure subjected to fire. A model that would consider the structure with the temperature distribution that is relevant at the required fire resistance time and would determine the load bearing capacity at that temperature would not be able to take the effects of unloading into account and would thus not comply with the requirement of the Eurocode.

If the material model recommended in the Eurocode is taken into account, effects of transient thermal creep need not be given explicit consideration. In fact, transient thermal creep, i.e. the additional deformation that occurs during first heating under load, is a feature known essentially for concrete. Steel is more prone to exhibit steady state creep, at least at elevated temperatures and load levels. This creep is supposed to be implicitly incorporated in the proposed stress-strain relationship and an explicit creep term is thus not required in the constitutive model.

Because numerical structural models can predict a very ductile behaviour of the member, it is not sufficient to let the computer make a run and simply note the fire resistance time. The deformations of the structure that have been calculated by the model have to be checked by the designer. It may happen, for example, that a simply supported steel beam exhibits at high temperatures a very large horizontal displacement of the end that is free, to a point that the beam would fall from its support in a real case. If the material model has an infinitely long plateau without a descending branch, it may even occur that the beam is totally folded by 180° in its centre, with one end coming to the other one and the load supported in pure tension. The displacements have also to be checked in order to verify that compatibility is maintained between all parts of the structure. The beam of a portal frame, for example, may not fall down below the level of the ground, something that would cause no problem to a computer program.

Any potential failure mode that is not covered by the model should be eliminated by appropriate means. Failure in shear or in local buckling are mentioned as example of such failure mode. This is correct only if a Bernoulli beam finite element is used. A model based on shell finite elements, for example, is perfectly able to track failure modes by shear and/or local buckling, but it has to be recognise that the utilisation of shell elements for modelling entire structures is not yet standard practice. The biggest danger when using a Bernoulli beam element is to bypass the verification of the class of the section because such an element treats all sections as Class 1, however thin the plates of the section are. Yet, contrary to a commonly heard opinion, hot rolled H and I sections are not systematically Class 1 or 2, especially in the fire situation.

What is an appropriate mean to eliminate other failure modes is not explained by the Eurocode. Shear can for example be checked separately by the simple calculation model and, if interaction between bending and shear has to be considered, it is possible to introduce a reduce yield strength for the elements near the supports where this is the case. The failure of a structure that comprises Class 2 elements should be specified by the user to occur when the first plastic hinge has developed; any further calculation that produces a redistribution of the bending moments between the sections along the elements should not be considered.

7.4.2 Capabilities of advanced mechanical models

The most important features of the advanced model are, first, the fact that *indirect effects of actions* are taken into consideration and, secondly, the fact that *large displacements* are taken into account exactly.

Thermal expansion of steel indeed generates various effects: a variation of the effects of actions due to restraint in statically indeterminate structures; a variation of stresses in the section due to temperature differences, see Figure 7.1, that exists even in statically determinate structures; large displacements that also modify the effects of actions because of *P-δ* effects. Whereas simple calculation models can take these effects into account only in an approximate manner, or not at all for some of them, the advanced model consider these effects exactly and continuously during the course of the fire. The bending moments, shear forces and axial forces vary continuously during the fire depending on the evolution of thermal expansion and of stiffness of the members and depending on the evolution of large displacements.

Figure 7.7, for example, shows the bending moment distribution in a multi-storey steel frame. The right hand extremity of each beam is fixed horizontally by a shear resisting concrete core that is not represented here. On the left is the moment distribution at time $t = 0$ and on the right is the moment distribution after 30 minutes of fire developing at the second floor. The axial compression force developing in the beam at the second floor, because of the restraint provided by the bending stiffness of the columns, and the tension force that develops in the beam at the third floor because of equilibrium reasons, highly modify the bending moment distribution during the course of the fire as illustrated in the right hand part in the Figure.

Figure 7.8 shows the evolution of forces and displacements in a quarter of a concrete slab supported on steel beams at the edges (Lim 2003). The deformed shape of the slab is shown before the fire and at failure on the left hand side of the Figure. The right hand side shows how the membrane force distribution is modified because of the large deflections. A compression ring develops at failure near the edges of the slab, allowing tension to develop in the centre of the slab. It is only when this membrane action mechanism is taken into account that the stability of the slab can be explained; any analysis that considers only bending as a possible load path fails to explain the high fire resistance times computed numerically and observed in experimental fire tests.

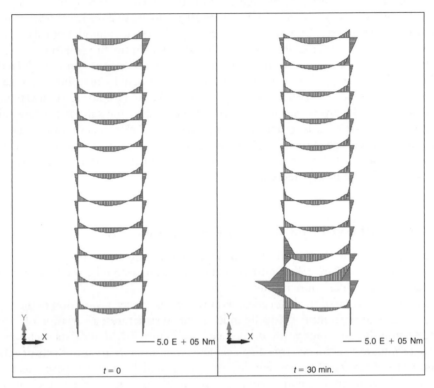

Fig. 7.7 Bending moment distribution in a multi-storey steel frame. *From Garlock & Quiel (2007)*

Theoretically speaking, the *physical dimensions* of the structure are not an issue in a numerical simulation, whereas experimental facilities are limited to dimensions in the order of a few meters. Larger and larger steel structures have been analysed in recent years, the destruction of the twin towers and of building 7 at the World Trade Center in New York City in 2001 being the most spectacular motivation in a common trend that existed anyway. Figure 7.9 shows the numerical model of a composite steel-concrete bridge with a span of 176 meters. This bridge suffered from a severe localized fire from a leaking gas pipe that it supported and collapsed in the water canal underneath.

Only with the advanced model is it possible to analyse *3D structures*. This possibility is normally not feasible through the simple calculation models, because of an excessive complexity of the problem, or to experimental testing, because of the size and of the complexity of the support and loading systems that would be required. Figure 7.10 shows a steel building failing partially because of a localized fire.

The *quantity and the quality of the results* obtained from a numerical analysis is higher by several order of magnitude than the results provided by simple calculation models or by experimental tests. It is as if several displacement and rotation transducers were present at every node of the mechanical model and there can easily be several hundreds of them. It is, on the contrary, not so easy to measure correctly the displacements of a structure located in a furnace and subjected to very high temperatures. Quantities such as strains, stresses or bending moments at virtually any point in

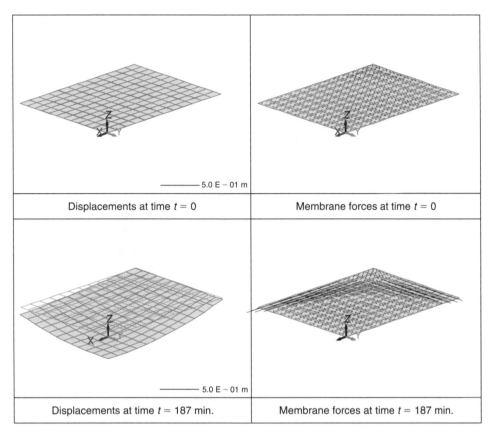

———— 5.0 E – 01 m	
Displacements at time $t = 0$	Membrane forces at time $t = 0$
———— 5.0 E – 01 m	
Displacements at time $t = 187$ min.	Membrane forces at time $t = 187$ min.

Fig. 7.8 Evolution of displacements and forces in ¼ of a concrete slab. *Courtesy Dr L. Lim, Univ. of Canterbury*

the structure are easily retrieved from a numerical analysis whereas they are not accessible from a test and are only approximated by the simple calculation models. The information that can be gained on the behaviour and the failure mode of the structure is extraordinary compared to the information gained from other methods that, very often, simply yield a fire resistance time.

7.4.3 *Limitations of the advanced mechanical models*

The size of the structure is, and probably will remain for a long time, one of the limits encountered in the step by step large displacement non-linear analyses that have to be performed in the fire situation. The towers of the World Trade Center are again the example that may come to mind first; others are the structure of the Piper Alpha oil platform that was destroyed by a fire following an explosion in the North Sea in 1988. A steel rack structure in a typical storage building analysed by two of the authors (Zaharia & Franssen 2002) contained an estimated number of 200 000 individual bars. The limit lies in the capabilities of the computer as well as in the resources, in term of

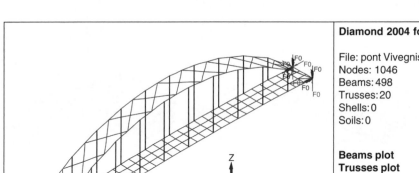

Diamond 2004 for SAFIR

File: pont Vivegnis
Nodes: 1046
Beams: 498
Trusses: 20
Shells: 0
Soils: 0

Beams plot
Trusses plot
Imposed DOF plot

▮ Deck.tem
▮ Arches.tem
▮ Ties.tem

Fig. 7.9 Composite steel-concrete bridge under fire

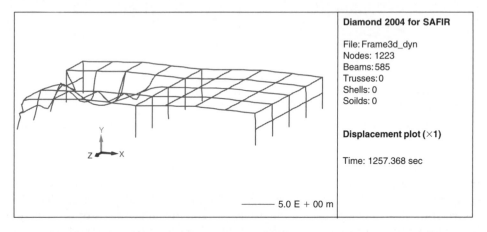

Diamond 2004 for SAFIR

File: Frame3d_dyn
Nodes: 1223
Beams: 585
Trusses: 0
Shells: 0
Soilds: 0

Displacement plot (×1)

Time: 1257.368 sec

—— 5.0 E + 00 m

Fig. 7.10 Failure of a 3D steel building due to localised fire

time and money, that can be allocated for descretising the structure, performing the calculations and analysing the results.

Because of the former limit, analysing a part of the entire structure, see Section 5.1, is very often necessary even when using an advanced calculation model. The choice of the forces or restraint to displacements that have to be applied at the *boundaries between the substructure* that is analyzed and the rest of the structure have to be chosen by the designer. No computer software will ever be able to substitute for the judgement of the designer. It has to be realized that the choice that is made can have a significant influence on the failure mode, certainly, and on the fire resistance time in most cases.

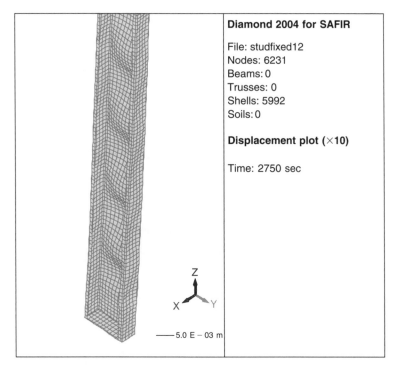

Diamond 2004 for SAFIR

File: studfixed12
Nodes: 6231
Beams: 0
Trusses: 0
Shells: 5992
Soils: 0

Displacement plot (×10)

Time: 2750 sec

—— 5.0 E – 03 m

Fig. 7.11 Steel stud wall in a fire analysis

A limit encountered by every numerical software that performs a step by step analysis of the structure is the too frequent occurrence of numerical false and premature failures. This problem could be traced to the impossibility for the software to cope with *local but temporary failures* that may occur in single elements or parts of the structure, whereas they do not endanger the load bearing capacity o the complete structure as a whole. Franssen and Gens (2004) showed that performing a dynamic analysis of the structure is a simple and effective way toward solving this problem.

It has to be realized that the *Bernoulli beam finite element*, certainly the workhorse of structural fire modelling during the last decade, has inherent limits that make it impossible to detect some failure modes such as, for example, local buckling and shear failure.

Figure 7.11 shows a steel stud wall to illustrate how local buckling can be tackled by using shell finite elements. Of course, although using shell elements is feasible for the analysis of a single member or of a detail, it is unthinkable for the analysis of a complete structure of significant size, at least for the time being.

Figure 7.12 shows the deformed shape of a cellular steel beam at failure. The length of the beam is short in this example, because this corresponds to a real specimen that has been tested experimentally. Note that the displacements have not been amplified in the drawing. Although the beam and the loading are normally symmetrical, there is a localization of the distortion near one end at failure. Such unsymmetrical failure

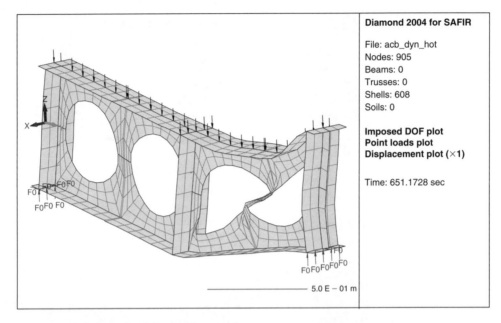

Diamond 2004 for SAFIR

File: acb_dyn_hot
Nodes: 905
Beams: 0
Trusses: 0
Shells: 608
Soils: 0

Imposed DOF plot
Point loads plot
Displacement plot (×1)

Time: 651.1728 sec

5.0 E – 01 m

Fig. 7.12   Failure of cellular steel beam in fire

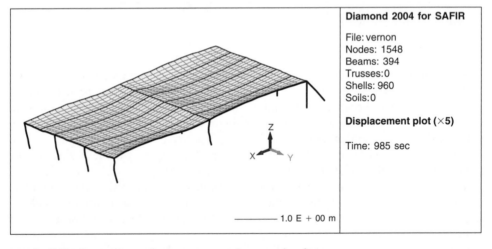

Diamond 2004 for SAFIR

File: vernon
Nodes: 1548
Beams: 394
Trusses:0
Shells: 960
Soils:0

Displacement plot (×5)

Time: 985 sec

1.0 E + 00 m

Fig. 7.13   Composite steel-concrete car park exposed to fire

modes are commonly observed in experimental tests. These very large displacements and the non-symmetrical failure mode could be obtained only with a dynamic analysis of the problem.

Beam and shell finite elements can of course be combined, as is the case for the car park presented on Figure 7.13. The steel columns and beams have been represented by beam elements whereas the concrete slab has been represented by shell elements.

7.4.4 Discrepancies with the simple calculation models

It has been explained that two correction factors, κ_1 and κ_2, are used in the simple calculation model for taking into account the effects of non-uniform temperature distribution, see Section 5.6.4.2.2.

It would theoretically be possible, in a numerical analysis, to make a full 3D thermal analysis of the beam, including the eventual protection and cooling effect induced at the supports. This is yet by far a too complex analysis for real applications and the effect is usually neglected. It has yet to be realized that, because of that, the results of an analysis by the advanced model may be less favourable than the results of an analysis of the same beam by the simple model. It is possible and easy to take the effect of colder temperatures at the supports in the advanced model in an approximate manner: a certain length of the elements on each side from the support may be given a less severe thermal axposure and, hence, a less severe temperature increase. What length has to be given to the affected zone and what level of decrease in severity of the thermal attack has to be chosen in order to achieve an increase of the load capacity up to $1/0.85 = 1.18$ is not known.

As already stated in Section 5.6.4.2.2, the increase to $1/0.70 = 1.43$ that is provided by the simple model for unprotected sections supporting a concrete slab will by no means be provided by the advanced model, except in the case of unsymmetrical sections, those with the top flange that is much weaker than the lower flange.

Chapter 8

Design Examples

8.1 General

This chapter presents four examples showing how a complete structure can be analised in fire situation using the concept of sub-structure analysis or element analysis.

The first example shows how a single span is extracted from a continuous beam allowing a member analysis by the simple calculation model. The discussion illustrates the concept explained in Section 5.1.2.

The second example illustrates the same concept for extracting a sub-structure from a moment resisting frame allowing analysis by the advanced model of the sub-structure instead of the complete structure.

The third example presents the steps to be followed when designing a hypothetical industrial building, using the simple method for the verification of the elements.

The last example presents a case study of a particular building: a high-rise rack steel structure supporting the building envelope. Due to the complexity of the structure, sub-structures have also to be defined, those being analyzed by an advanced model. The discussion here encompasses also the decisions that were taken for representing the fire exposure.

8.2 Continuous beam

The procedure described in Section 5.1.2 is applied here for a continuous beam that is not restrained against axial expansion, see Figure 8.1. The mains steps to be followed in the analysis are as follows:

Step 1   Determine the effects of action in the whole beam under the design load combination for the fire situation. In this particular case, this step might be omitted as will be seen later in step 5.

Step 2   Each span will be successively analysed as an element. This is because, on one hand, the structural analysis of a single span beam is a trivial problem with very well known solutions, whatever the boundary conditions, and, on the other hand, the plastic theory indicates that the ultimate load of a span is not affected by the loads applied on the other spans. Here, the load is constant, but the same theory says that the fire resistance of a span is not influenced by the loads applied on the other spans. The next steps here deal with the analysis of the central span.

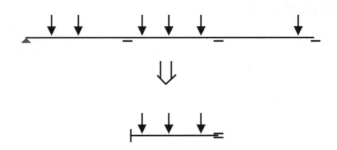

Fig. 8.1 Continuous beam exposed to fire

Step 3   The vertical restrains provided by the supports at both ends of the span are taken into account for the element because these are supports of the structure.
Step 4   The loads applied on this central span are taken into account.
Step 5   Degrees of freedom at the boundary.
  (a)   For the horizontal degrees of freedom, one is fixed, in order to prevent a rigid body movement, and on the other one the displacement is not fixed but the axial force determined during the structural analysis of the total beam is applied, in this case 0. This way, no axial restraint is created in the element, which corresponds to the situation in the structure.
  (b)   The rotational degrees of freedom are fixed (this would not be the case for the exterior supports of first and last span). The fact that they are fixed will allow the development of the plastic hinges on the supports, in similarity with the failure mode that develops when this span fails in the structure. Of course, plastic hinges can develop only in Class 1 and Class 2 sections.
Step 6   The effects of actions, here bending moments and shear forces, can be determined in this element, a one span beam with fixed end rotations.
Step 7   No indirect fire action is considered in an element analysis.

The resistance and stability of this span of the beam is then verified according the appropriate equations from Section 5.6.4 or 5.6.6 taking into account that the beam is laterally restrained or not, that the temperature in the section and along the beam is uniform or not, and depending whether the section is a Class 1, 2, 3 or 4 section in the fire situation. The procedure has to be repeated for every span, every load case and every fire scenario.

8.3   Multi-storey moment resisting frame

The procedure described in section 5.1.2 is applied here for a multi-storey moment resisting frame, see Figure 8.2. This example shows how a sub-structure is extracted from the frame and gives the considerations at the base of the choice for the boundary conditions between the sub-structure and the rest of the frame. It is assumed that each floor of the building constitutes a separate fire compartment.

Step 1   Determine the effects of action in the total structure under the fire load combination that is being considered.

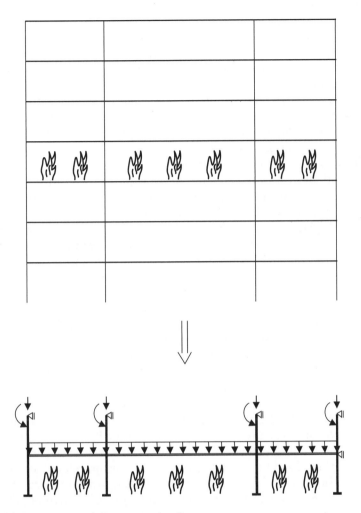

Fig. 8.2 Moment resisting frame exposed to fire

Step 2   The first decision, concerning the limits of the substructure, is here to limit the analysis to a plane frame. This is usually the case if the structure of the entire building is made of parallel similar frames with only secondary beams spanning from one frame to the other. In this plane frame, the columns directly exposed to the fire have to be part of the substructure, as well as the beam which is directly exposed to the action of the fire, i.e. the beam above the fire compartment. The beam and columns under the fire compartment may be excluded, because they will be represented by rigid boundary conditions, see next step. The columns above the fire compartment, on the other hand, will be included in the substructure because they offer less restraint to rotation at the top of the fire columns and their flexibility will influence the moment redistribution in the part of the substructure that is directly affected by the fire.

Note: in this case, and strictly speaking, there is an interaction between steps 1 and 2. The true decision process would probably be first to decide to make a 2D plane analysis (and this is a decision that belongs to step 2), then to make step 1, the analysis of the structure, i.e. the plane frame only.

Step 3   The substructure shown in the lower part of Figure 8.2 has no direct support to the foundations.

Step 4   The loads directly applied on the substructure are taken into account. In this case they consist mainly of the dead weight and of the loads applied on the beam above the fire compartment. Wind load could also be applied on the external columns if relevant.

Step 5   Degrees of freedom at the boundary.

(a)   All degrees of freedom out of the plane of the substructure are fixed at each point where a cut has been made to separate the substructure out of the structure, i.e. certainly at all beam to column joints. No out of plane displacement or rotation can exist at these points. Out of plane displacements will probably also be fixed for the beam, owing to the presence of a floor slab. Whether out of plane displacements will be fixed for the columns depends on the local arrangements of the building. This will be the case, for example, if there are transverse masonry walls between the columns. In this case, the substructure will be considered as a 2D object in the structural analysis. If this is not the case, a 3D analysis may prove to be necessary, even if the substructure is plane, in order to be able to check the out of plane stability of the columns. If the frame is a sway frame, it would be more representative of the real behaviour not to fix the horizontal displacements at the top of the columns but to impose the same displacement for all the columns.

(b)   If the frame is a braced frame, it will be assumed for the analysis of this substructure that the efficiency of the bracing system is maintained for a duration that is longer than the fire resistance of the substructure. Three fixed horizontal restraints are thus introduced at the connection between the bracing system and the substructure, here at the bottom, at mid level and at the top of the right hand column. The fire resistance of the bracing system has to be checked in a subsequent analysis, considering the bracing system as another substructure.

(c)   At the bottom of the heated columns,

- the vertical displacement is fixed. This reflects the fact that the true vertical displacements at these points are very similar for all the columns,
- the horizontal displacement of all columns will be the same, which reflects the high axial stiffness of the lower beam. Because the bottom of the right hand column is fixed by the bracing system, all columns will be fixed,
- the rotation is also fixed, owing to the presence at each joint of the lower beam and the lower column which remain cold and thus much stiffer than the heated columns of the fire compartment.

(d)   At the top of each non-heated column,

- the vertical displacement is free, because there is no reason to introduce an axial restrain at these points. The axial force resulting from the analysis of the entire structure at time $t = 0$ is applied here,

- the horizontal displacement of all columns is the same, owing to the axial stiffness of the upper beam. Because the right hand column is fixed by the bracing system, all columns will be fixed.
- the rotation is free and the bending moment derived from the analysis of the entire structure is applied.

Step 6   It is not necessary to determine the effects of actions at time $t = 0$, see step 7.

Step 7   Indirect fire actions have to be considered in the substructure which implies that, practically, the only possible calculation method is the advanced calculation model, see Section 5.2.1 and Chapter 7. The effects of actions are thus evaluated continuously during the course of the fire and their value at time $t = 0$ is not relevant.

The procedure is repeated for every load case and every fire scenario, i.e., in case of a nominal fire, for the fire being supposed to be located successively at each different floor.

8.4   Single storey industrial building

The structure considered in this example is presented on Figure 8.3.

The skin of the building is made of corrugated steel sheets. The diaphragm effect from the sheeting is nevertheless not considered in the design of the building, at least not in the fire situation because of the uncertainty on the behaviour of the connectors at elevated temperature. Seam fasteners, for example, may well be made of aluminium alloy blind rivets and aluminium looses strength faster than steel under increasing temperatures. Steel self-drilling self-taping screws used as sheet to perpendicular member fasteners may be associated with neoprene fasteners and neoprene may soften under high temperatures.

The sheets of the roof and of the walls are supported by continuous steel purlins spanning from frame to frame.

The transversal frames, with hinged bases, are linked by means of rafters at the level of knee and corner joints.

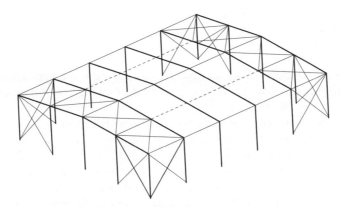

Fig. 8.3 Steel industrial building exposed to fire

Intermediate columns supporting the vertical loads as well as the wall purlins are placed in the two end frames.

Cross-bracings are provided between the frames in the two end spans, in the lateral walls and in the roof. In the transversal direction, cross-bracings are provided in the lateral parts of end frames.

The loads to be considered are:

- The permanent load, G
- The wind in the transversal direction, WT
- The wind in the longitudinal direction, WL
- The snow load, S

If the frequent value of the dominant variable action is taken into account, see Equation 2.5.a, the load combinations in case of fire are:

- Combination 1: $G_k + \psi_1 S_k + \psi_2 WT_k$
- Combination 2: $G_k + \psi_1 S_k + \psi_2 WL_k$
- Combination 3: $G_k + \psi_1 WT_k + \psi_2 S_k$
- Combination 4: $G_k + \psi_1 WL_k + \psi_2 S_k$

If the quasi-permanent value of the dominant variable action is taken into account, see Equation 2.5.b, the load combinations in case of fire are:

- Combination 1: $G_k + \psi_2 WT_k + \psi_2 S_k,$
- Combination 2: $G_k + \psi_2 WL_k + \psi_2 S_k.$

The effects of actions in the structural elements have to be determined in fire situation, at time $t = 0$. The analysis of the structure is performed in the same way as for the analysis at room temperature, considering successively either the four or the two load combinations presented above.

An appropriate fire scenario has to be chosen. Considering that an analysis by a CFD model will probably not be performed for such a simple and relatively common structure, the following fire models may be considered:

- the standard ISO temperature-time curve,
- localised fire models,
- two zone models.

The parametric temperature-time curve may be applied if the dimensions of the fire compartment (considered as the entire volume of the building) do not exceed the limits of application of this simplified fire model.

Thermal actions from localised fires may be taken into account until the flash-over occurs. The occurrence of the flash-over may be determined by a simulation of the fire development, using an advanced two-zone model. If for the required fire resistance time of the main structural elements the flash-over is still not produced, then it is sufficient to perform the analysis of different elements under combined effects of the localised fire and of the hot zone of the two zone model. If the zone model indicates the occurrence of flash-over, the analysis should be continued from that time and further in the one-zone condition, without any effect from the localised fire.

Note: the software OZone developed at the University of Liege, Cadorin and Franssen 2003 and Cadorin et al. 2003, has the ability to switch automatically from a two zone to a one zone configuration when certain criteria are met. It can be downloaded for free from the web site of the university.

By default, in case of lack of the appropriate software tool to determine a fire scenario considering a two-zone advanced model, the standard ISO temperature-time curve is always accepted.

For this example, all further considerations concerning the analysis of the elements in case of fire are made assuming that the entire compartment is engulfed in fire and the temperature of the gases in the compartment follows the standard ISO temperature-time evolution.

The normal and most usual situation is that the required fire resistance time is known, since it is prescribed by the relevant fire authorities or in the local building codes. The analysis of each structural element can thus be performed in the load domain.

The required fire resistance of the frames is verified on the basis of the governing equations given in Chapter 5 considering each separate element, namely the columns, subjected to combined bending moment and axial force, and the beam, subjected mainly to bending moment. The fire resistance of the frame is the lowest resistance of all the elements. It comes then as a consequence that, if the required fire resistance for beams and columns is different as is the case in some regulations, function of the building destination, it is indeed meaningless in the case of single storey frames like in the present example. Even if the required fire resistance for beams is less than that of the columns, the same fire resistance time will in fact be obtained for all elements, because collapse of the beam produces the collapse of the entire frame.

The calculation of the fire resistance of the frame, i.e. of its elements, columns and beam, is performed assuming that the stability of all other members that are required for the stability of the frame is ensured. These are, for example, the bracings and the rafters. Of course, the fire resistance of these elements will have to be checked in a subsequent stage of the analysis. Indeed, the columns may be considered as fixed in the longitudinal direction by the rafters at the corner connection only on the condition that the rafters and the cross-bracings from longitudinal walls are still effective, at least until the corresponding required time resistance. Therefore, all the cross-bracings and rafters in the longitudinal walls must be designed to withstand the loads in the fire situation resulting from the static analysis, during the same time as the columns. The buckling length of the columns to be considered in this case for the longitudinal direction would be the same as for room temperature, i.e. the column height. As a simplification, the buckling length in the plane of the frame may also be calculated as for room temperature.

The beam may be considered laterally fixed between the corner and the knee connection of the transverse frame, provided that the conditions above are fulfilled and also the rafters at knee joint and the bracing system in the roof are effective at the corresponding resistance time of the beam. Considering the lateral restraint of the beam against lateral-torsional buckling, this could be provided for the upper compressed flange by the purlins. But the problem is that, generally, the roof is considered as a secondary system of the structure. It is thus not required to have any fire resistance time. The purlins can perhaps stabilise the beam if they are able to act as tension elements transmitting the stabilising effect to one of the end bracing systems. Two end bracing systems have to be present if this mechanism is to be expected to work because the lateral instability of the beam may be in either direction. If this is not the case, and if the beam does not satisfy the required resistance time, instead of reinforcing the purlins until they can work in compression, which could lead to important supplementary costs, additional rafters may be provided in the plane of the roof, at mid span of the beam, in line with the front columns, as marked on Figure 8.3.

The end frames have to be verified separately. It is indeed likely that the presence of the columns in the end gables has been accounted for in the design at room temperature and that the members of these frames are somewhat weaker than the design of the current frames.

For the front columns in the end gables, the buckling length may be considered equal to that determined at room temperature, i.e. equal to the length of the columns for both directions. Being part of the front transverse frame, the front columns are subjected to combined axial force, received from the end transverse frame, and to bending moment from wind action. Their resistance time in fire conditions must be the same as that of the columns and beams of the frames, considering that their collapse may produce the collapse of the end transverse frames.

Due to the presence of the hinged columns in the end gables, which offer supplementary support to the beams but no rigidity to the lateral loading, and because the end frames take less vertical loading than the main transverse frames, the lateral columns and the beams of these end frames are weaker that those of the main transverse frames. Therefore, the vertical cross-bracings in the end gables will be designed in order to ensure for the front transverse frames at least the same stiffness as that of the main transverse frames. Because no precise recommendation is given for limiting the displacements in case of fire (see Section 1.2), it may be considered that the resistance time of these cross-bracings is not critical. However, if the resistance time of the cross-bracings in the front transverse frame is less than the required resistance time, a supplementary verification must be performed for the front frames, considering that the cross-bracings are not present. It must be noted that, considering the flexibility of the front frame in this case, second order effects may became important and should then be considered in the static analysis, which is not easy with the simple calculation model.

The bracings that resist the longitudinal wind loads have to be verified. Usually, only the diagonals that work in tension are considered, while those that are submitted to compression are neglected because their high slenderness induces buckling failure. Diagonals that work in compression can be considered if they are appropriately sized.

Finally, the rafters that stabilize the frames have also to be checked. The axial forces required to mobilise the stabilising effect to the frame have to be evaluated. It is likely that the load combination with the longitudinal wind loads will be the most critical one for these elements.

The high number of elements to be verified, combined with the number of load combinations (consider that each wind load may in fact comprise various possibilities for the inside pressure), lead to an enormous amount of calculations to be performed. It has also to be considered that the procedure described above is only applicable for verifying a structure with known dimensions and properties. In the case that the outcome of the fire verification is not satisfactory, a new verification has to be performed with new sections.

This is why the advanced calculation models, although they required an additional cost for acquiring the appropriate software and for learning to use it, offer enormous advantages. Most often, the advanced model would be used to analyse one frame as a 2D statically indeterminate structure loaded in its plane but modelled by 3D beam elements in order to allow out of plane instability to develop. Other elements such as rafters and members of the bracing system would be analysed by the simple calculation model because the mechanical behaviour of these elements is much simpler. It may be possible that in the future the whole building structure will be modelled completely and analysed as one single object, as is more and more often performed at room temperature. The main difficulty in such a global analysis is the fact that a great number of members exhibit instability because of the restraint forces induced by thermal expansion. A step by step algorithm relying on successive static analyses is not suitable for analysing such a complex structure. A dynamic analysis is normally required, see Section 7.3.3.

8.5 Storage building

This example presents a fire design case study for a high-rise storage rack system that supports the skin of the building. From the legal point of view, the classic rack systems located within a building with its own independent structure are usually not required to have any fire resistance. These racks are considered as furniture. On the contrary, if the skin of the building is directly applied on the rack system, the racks then become the structure of the building, and they must have the required fire resistance.

The structure considered here (Zaharia & Franssen, 2002) is a steel storage building of this type built by the Belgian company TRAVHYDRO. It has a floor surface of $9168\,m^2$ for 30 m high. There are 36 racks on the 160 m length of the warehouse. Between the cross-aisle frames of the racks, horizontal elements are provided in order to maintain the distance between the rails for the wagons of the automatic pallet transport system, as shown on Figure 8.4. On the down-aisle direction, one rack has 60 m length and is provided with 10 levels for pallet disposal, see Figure 8.5.

For this type of industrial building in which hardly any person is present, and accounting for the existence of an automatic sprinkler extinguishing system, a fire resistance time of 15 minutes is required. Concerning the safety of the occupants of the building, it must be emphasised that only two people are authorised to enter the building, for two hours, once a week. These people are trained for fire situations, know the building, and they may evacuate in less than 10 minutes from the moment of fire discovery. A special requirement for this building was that, even if one rack of the structure collapses, because of a malfunction of the sprinkler system for example, the entire building must not present a progressive collapse phenomenon. This requirement is for preserving the safety of fire fighters that may have entered the building to fight the fire that engulfs one single rack.

The analysis of the warehouse subjected to fire was performed with the SAFIR computer program.

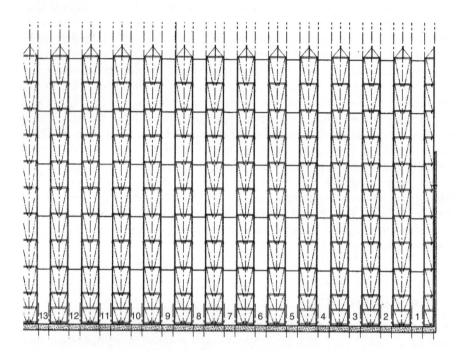

Fig. 8.4 Cross-aisle direction (elevation) in a warehouse

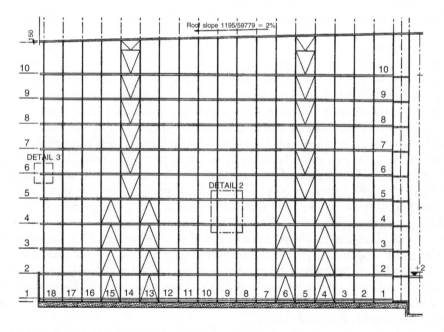

Fig. 8.5 Down-aisle direction (elevation) in a warehouse

A mechanical analysis under elevated temperatures considering the standard ISO fire in the entire compartment was realised first, and showed that the 15 minutes of fire resistance required for this type of building cannot be obtained. The reasons are:

- First, the low thermal massivity of the cold-formed profiles normally used to build rack systems leads to a rapid temperature increase in the members and, consequently, a rapid decrease of the load bearing capacity.
- A second reason is that the combination factors for frequent or quasi-permanent variable actions for storage systems are much higher than for other types of buildings, typically in the order of 0.8 to 0.9, see Table 2.1. The structure is thus highly loaded in the fire situation.
- Also playing a key role are the important indirect effects of action caused by thermal expansion in such building systems, often continuous over 100 meters, and supposed to be uniformly heated when submitted to a nominal fire.

It is thus desirable to consider a more realistic approach for the fire scenario, taking into account the pre-flashover phase as well as different physical parameters, concerning the fire load and the building itself. In order to determine the temperature-time curves for the thermal analysis of cross-section, using the combined two zone and one zone model, the computer program OZone developed at the University of Liege was used. Depending on the fire spread area, of the height of the hot zone or on the temperature in the hot zone, a transition from the two zone situation to a one zone situation, with an uniform temperature in the entire compartment, may occur. Flashover is assumed to occur when the temperature in the hot zone reaches 500°C.

There are no large openings in the storage building. There are no windows, only normal doors for personnel access. Smoke exhaust systems are provided in the ceiling, with a surface equal to 2% of the surface. The fire load density was evaluated as 8400 MJ/m^2, taking into account the combustibles provided by the stored goods on the 25 m height of storage in the building, and the maximum rate of heat release density is 8620 kW/m^2, according to a previous study made by CTICM (Joyeux & Zhao, 1999).

A very important parameter for the evolution of the temperature-time curve is the fire growth rate. Recent tests on rack fires made at TNO in The Netherlands demonstrated that the fire growth rate in racks may be faster than *"ultrafast"*, to which corresponds a time constant of 75 seconds in a t^2 model. For this building, a time constant of 66 seconds was taken into account as the time to obtain 1 MW of heat release, which represents the fastest fire growth rate obtained in the experimental tests.

On the basis of these hypotheses, the temperature-time curve for the hot zone obtained with the two zone model is shown in Figure 8.6. The compartment is under a two zone situation until 17 minutes after the fire beginning, when the interface of the two zones reaches less than 20% of the building height, so one of the criteria for one zone transition is fulfilled. The flashover is reached after 24 minutes.

After 10 minutes of fire, the thickness of the smoke layer is around 20 meters and the temperature in the lower zone is less than 30°C, so the smoke and the temperature should not impair the ability of the occupants to evacuate.

For the rack where the fire starts, the fire scenario supposes that a column and the adjoining rails are subjected to localised fire. The rest of the structure is under

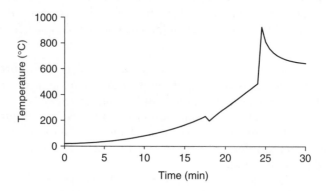

Fig. 8.6 Temperature-time curve for the upper zone in the warehouse

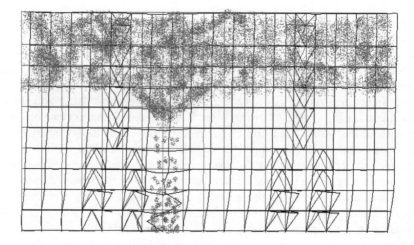

Fig. 8.7 Collapse of one rack under natural fire

the influence of the two zone temperature distribution, as indicated schematically by Figure 8.7.

Because it is totally unrealistic to model the whole structure in the mechanical analysis, a series of uncoupled analyses is performed considering separately the behaviour in one plane, Figure 8.4, then in the other planes, Figure 8.5.

Because of the high restraint to thermal expansion in this highly indeterminate structure, the buckling of some elements within the bracing system of the rack occurs after less than 3 minutes of fire exposure, but this does not lead to the collapse of the entire rack. The numerical analysis indicates that the global collapse of the rack under fire occurs after around 6 minutes, just after the collapse of the upright under localised fire.

In the other direction, and assuming that the fire started in the rack located in the middle of the building, it can be considered that the building is cut in two separate parts by the collapse of the rack under fire at 6 minutes, as shown in Figure 8.8. After the collapse of the first rack, it can be considered that the fire spreads to the two adjoining

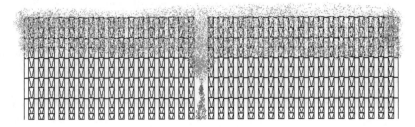

Fig. 8.8 Cross-aisle configuration after collapse of the first rack

frames, and they are under the influence of a localised fire. The working hypothesis is that the collapse of a rack produces the breaking of the horizontal elements between cross-aisle frames so it does not produce the collapse of adjoining frames. Therefore, in order to ensure that the progressive collapse is avoided, it is necessary to design the horizontal elements connections to withstand the necessary compression and tension efforts induced by the wind and by the movement of the wagons of automatic trans-port system, but to have poor resistance for relative displacements between frames in both vertical and horizontal directions. The proper design of these elements and their relevant connections that ensure the appropriate behaviour may be the biggest challenge for the fire safety of this structure.

These horizontal elements between the cross-aisle frames are the next elements that will suffer and fail from the fire after the collapse of the first rack, especially those that are located in the upper zone, see Figure 8.8. Indeed, due to the continuous temperature increase in the upper zone, these elements exhibit thermal expansion and the horizontal displacement that should be produced at the top of the racks is opposed to by the stiffness of the lower and colder part of the racks. The compression forces in the horizontal elements increase until failure by buckling occurs, usually in the middle of the remaining structures, i.e. in the case shown on Figure 8.8 at ¼ and ¾ of the length of the hall. This buckling reduces the restraint forces in the elements, for a while at least because further temperature increase will again induce increasing compression forces in the elements that have not yet failed. The analysis of the cross-aisle direction of the building, see Figure 8.8 is thus performed, considering the expansion and the progressive suppression of the horizontal elements. The analysis using the two zone situation for the entire building combined with the localised fire for the racks in the middle of the building shows that collapse occurs within about another 6 minutes in the two racks under localised fire.

The localised collapse of the racks in the middle of the building may continue in the same manner, without producing progressive collapse of the whole structure, on the condition that the horizontal elements between adjacent racks break at the moment of collapse of one of the racks that they connect.

The global collapse of the entire structure corresponds to the moment of collapse of all other current racks, not under localised fire, but under the influence of the increasing temperature in the two zones.

The structure as originally proposed exhibited a global and progressive collapse before flash-over occurs and this situation was judged as unacceptable because this could be a great thread, or even a possible killer, for fire fighters that may have entered the building owing to the fact that the situation is still tenable. Some slight modifications have been proposed, mainly in the bracing system, until it could be shown that the modified structure survives long enough to reach the time corresponding to the flashover phase. After this limit, it would be difficult to obtain more structural resistance, taking into account the high level of temperatures in the compartment. It would also probably be totally irrelevant.

The collapse of the first racks under localised fire, after 6 or 12 minutes, does not present a danger because at this moment the persons who evacuate the building cannot be present in the area that is engulfed in fire.

Considering the fire fighting and rescue services, which may arrive in less than 15 minutes after the fire beginning, the main risk for those who may enter the building is the progressive collapse of the structure. The present study tends to indicate that no progressive collapse has to be feared as long as the flash over does not occur. When and if this happens, the structure is certainly in great danger of immediate collapse but, at this time, this is not anymore a real issue for the firemen, who should have evacuated the compartment.

There is certainly an uncertainty attached to the results presented because of the uncertainties linked to the data and to the fire model. Such a study can nevertheless provide very interesting and useful information. It shows that it is nearly impossible to strengthen the structure until it presents a fire resistance of 15 minutes to the standard fire. Trying to reach this objective would cost enormous amount of money and would not necessarily improve the fire safety in the building. Indeed, because any real fire would start as a localized fire, it may even occur that the additional stiffness induced in the system would be detrimental to the stability of the rack under fire.

Analysis of the development of the temperatures likely to exit in case of a fire shows that there is a sufficient period of time available for evacuation before flash-over. This is because of the huge volume of the compartment in which the energy produced by the fire is dissipated as well as to the presence of heat and smoke evacuation systems in the roof.

The failure of the racks under fire cannot be avoided, but can probably be accepted. The key issue is that this local failure does not lead to progressive collapse of the whole structure. The failure mechanism has been identified and the role of the horizontal elements between the racks has been pointed out. Care should be taken in order to ensure an appropriate behaviour of these elements.

Simple and economic modifications of the structure have been proposed in order to extend the time before global collapse, the aim being that it does not occur before flash-over occurs and makes the situation in the compartment untenable for any occupant eventually still in the building or for fire fighters who may have entered the building.

High Temperature Properties and Temperature Profiles

I.1 Thermal properties of carbon steel

The following sections describe the thermal properties of carbon steel. The thermal properties of stainless steel are given in Annex C of Eurocode 3 and in Appendix of ASCE manual (1992). However, AISC steel design manual lists Eurocode thermal properties. There are slight differences between Eurocode and ASCE specified high temperature thermal properties.

I.1.1 Eurocode properties

I.1.1.1 Thermal conductivity

Thermal conductivity of carbon steel decreases with temperature in the manner described by Eq. I.1, see Figure I.1.

$$\lambda_a = 54 - \theta_a/30 \geq 27.333 \tag{I.1}$$

where θ_a is the steel temperature in °C.

Although this is not explicitly stated in Eurocode 3, the thermal conductivity of steel is generally assumed to be reversible during cooling, which means that the thermal conductivity of steel varies according to Eq. I.1 during heating from 20°C to a temperature $\theta_{a,max}$ as well as during subsequent cooling back to 20°C.

I.1.1.2 Specific heat

Specific heat of carbon steel in J/kgK is varying with temperature in the manner described by Eq. I.2, see Figure I.2 (in kJ/kgK).

$$
\begin{aligned}
c_a &= 425 + 0.773\,\vartheta_a - 1.69 \times 10^{-3}\,\vartheta_a^2 \\
&\quad + 2.22 \times 10^{-6}\,\vartheta_a^3 & \text{for} \quad \vartheta_a < 600°\text{C} \\[4pt]
c_a &= 666 + \frac{13\,002}{738 - \vartheta_a} & \text{for} \quad 600°\text{C} \leq \vartheta_a < 735°\text{C} \\[4pt]
c_a &= 545 + \frac{17\,820}{\vartheta_a - 731} & \text{for} \quad 735°\text{C} \leq \vartheta_a < 900°\text{C} \\[4pt]
c_a &= 650 & \text{for} \quad 900°\text{C} \leq \vartheta_a
\end{aligned}
\tag{I.2}
$$

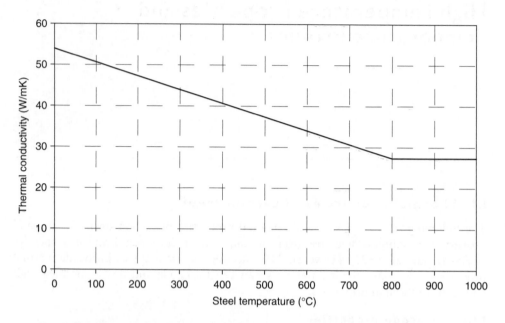

Fig. I.1 Thermal conductivity of carbon steel

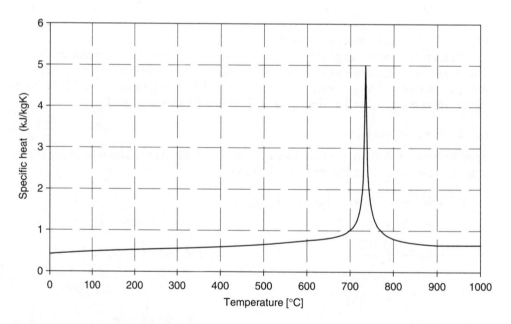

Fig. I.2 Specific heat of carbon steel

The peak in the curve around 735°C is due to the crystallographic phase change of the material. This peak will induce an S shape in the curves that show the evolution of steel sections with time; the temperature increase is slowing down around 735°C, just to accelerate again for higher temperatures.

I.1.2 Thermal properties of steel according to ASCE

I.1.2.1 Thermal conductivity

ASCE (Lie, 1992) provides empirical relationships for high temperature thermal properties of steel. While the trends in thermal conductivity are similar to that of Eurocode, there are slight variations in the actual values at different temperature ranges.

$$for \ 0 \le T \le 900°C \quad k_s = -0.022T + 48 \ Wm^{-1}\,°C^{-1} \tag{I.3}$$

$$for \ T > 900°C \quad k_s = 28.2 \ Wm^{-1}\,°C^{-1} \tag{I.4}$$

Where T is temperature in steel and
k_s is thermal conductivity.

I.1.2.2 Specific heat

ASCE lists thermal capacity relationships for structural steel at various temperature ranges. The thermal capacity, defined as product of specific heat and density, trends are similar to those of Eurocode specific heat values, but there are slight variations in actual values at different temperature ranges.

$$for \ 0 \le T \le 650°C$$
$$\rho_S C_S = (0.004T + 3.3) \times 10^6 \ Jm^{-3}\,°C^{-1} \tag{I.5}$$
$$for \ 650°C < T \le 725°C$$
$$\rho_S C_S = (0.068T + 38.3) \times 10^6 \ Jm^{-3}\,°C^{-1}$$
$$for \ 725°C < T \le 800°C$$
$$\rho_S C_S = (-0.086T + 73.35) \times 10^6 \ Jm^{-3}\,°C^{-1}$$
$$for \ T > 800°C$$
$$\rho_S C_S = 4.55 \times 10^6 \ Jm^{-3}\,°C^{-1} \tag{I.6}$$

Where T is temperature
ρ_s is density
Cs   is specific heat

I.2 Thermal properties of fire protection materials

The room temperature thermal properties of commonly used fire protection materials are given in ECCS (1995) and are reproduced in Table I.1. The data on high temperature properties for many of these materials are scarce. The general trends for some of these high temperature properties can be found in SFPE handbook [Kodur and Harmathy 2002].

Table 1.1 Thermal properties of commonly used fire protection material (ECCS 1995)

Material	unit mass ρ_p [kg/m³]	mosture content p [%]	thermal conductivity λ_p [w/(m · k)]	specific heat c_p [J/(kg · K)]
Sprays				
– mineral fibre	300	1	0.12	1200
– vermiculite cement	350	15	0.12	1200
– perlite	350	15	0.12	1200
High-density sprays				
– vermiculite (or perlite) and cement	550	15	0.12	1100
– vermiculite (or perlite) and gypsum	650	15	0.12	1100
Boards				
– vermiculite (or perlite) and cement	800	15	0.20	1200
– fibre-silicate or fibre-calcium-silicate	600	3	0.15	1200
– fibre-cement	800	5	0.15	1200
– gypsum board	800	20	0.20	1700
Compressed fibre boards				
– fibre silicate, mineral-wool, stone-wool	150	2	0.20	1200
Concrete	2300	4	1.60	1000
Light weight concrete	1600	5	0.80	840
Concrete bricks	2200	8	1.00	1200
Bricks with holes	1000	–	0.40	1200
Solid bricks	2000	–	1.20	1200

The mineral fiber mixture combines the fibers, mineral binders (usually, Portland cement based), air and water. It is a limited combustible material and a poor conductor of heat. Mineral fiber fire protection material is spray-applied with specifically designed equipment which feeds the dry mixture of mineral fibers and various binding agents to a spray nozzle, where water is added to the mixture as it is sprayed on the surface to be protected. In the final cured form, the mineral fiber coating is usually lightweight, essentially non-combustible, chemically inert and a poor conductor of heat.

I.3 Temperatures in unprotected steel sections (Eurocode properties)

Table I.2 Temperature in unprotected sections subjected to the ISO fire

| A_m^*/V [m^{-1}] | 400 | 200 | 100 | 60 | 40 | 25 |
V/A_m^* [mm]	2.5	5.0	10.0	16.7	25.0	40.0
Time [min.]			Steel temperature in °C			
0	20	20	20	20	20	20
5	430	291	177	121	90	65
10	640	552	392	276	204	142
11	661	587	432	308	228	159
12	678	616	469	340	253	177
13	693	642	503	371	278	194
14	705	663	535	402	303	212
15	716	682	565	432	328	230
16	725	698	591	460	353	249
17	732	711	616	487	377	267
18	736	721	638	513	401	286
19	743	729	658	538	425	304
20	754	734	676	561	447	323
21	767	738	692	583	470	341
22	780	744	706	604	491	360
23	790	754	717	623	512	378
24	799	767	726	641	532	396
25	807	780	732	658	551	414
26	813	792	735	674	570	431
27	820	803	740	688	588	449
28	826	813	746	701	604	466
29	831	821	756	712	621	482
30	837	828	767	721	636	498
31	842	835	780	728	651	514
32	847	841	793	733	665	530
33	852	846	805	736	678	545
34	856	851	816	740	690	559
35	861	856	827	745	701	573
36	865	861	836	753	711	587
37	870	866	844	763	719	601
38	874	870	852	774	726	614
39	878	874	859	786	731	626
40	882	878	865	798	734	638
45	900	897	890	852	761	692

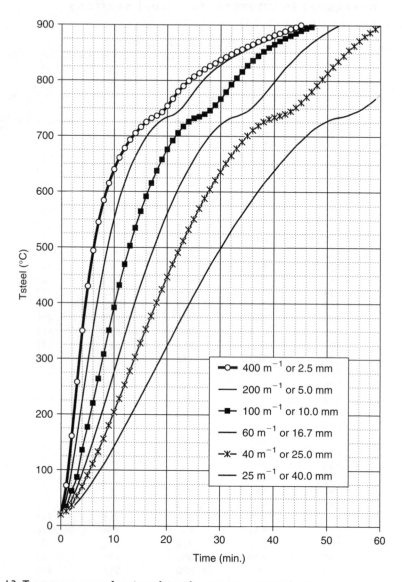

Fig. I.3 Temperatures as a function of time for various massivity factors

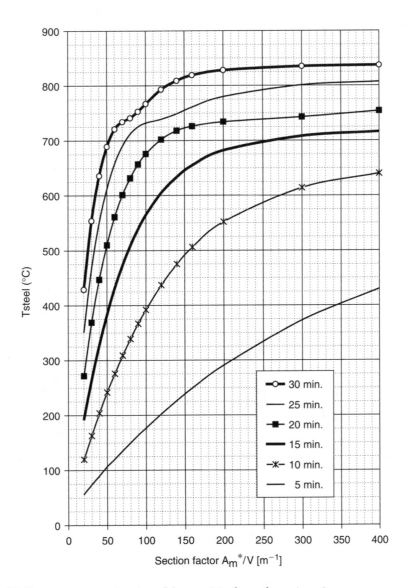

Fig. I.4 Temperatures as a function of the massivity factor for various times

1.4 Temperatures in protected steel sections (Eurocode properties)

Table 1.3 Temperature in protected sections subjected to the ISO fire

k_p [W/m^{3}K]	200	400	600	800	1200	2000
Time [min.]			Steel temperature in °C			
0	20	20	20	20	20	20
10	37	54	70	85	113	163
20	60	97	130	160	215	304
30	84	139	188	232	306	421
40	108	181	244	298	388	514
50	132	222	296	359	459	589
60	156	260	345	414	520	650
70	179	298	391	465	573	699
80	202	333	433	510	620	730
90	225	367	472	552	661	743
100	247	399	509	589	695	773
110	268	430	542	623	721	816
120	289	459	573	654	734	859
130	310	486	602	681	744	900
140	330	512	629	705	765	935
150	349	537	654	723	795	965
160	368	560	677	733	828	990
170	386	582	697	739	861	1013
180	404	603	714	751	892	1032
190	422	623	727	769	921	1049
200	439	642	734	792	948	1065
210	455	660	738	817	972	1078
220	471	677	747	843	993	1090
230	487	692	760	869	1013	1101
240	502	706	777	893	1031	1112

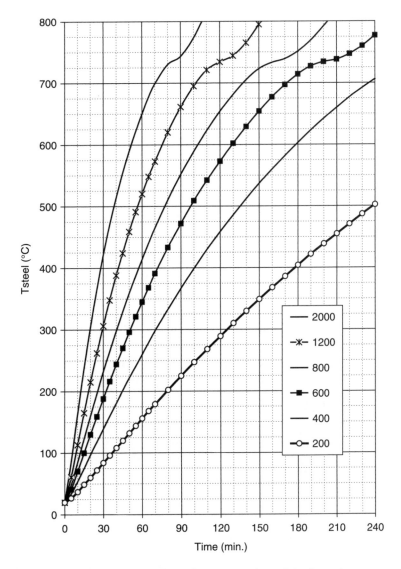

Fig. I.5 Temperatures as a function of time for various values of the factor k_p

Mechanical Properties of Carbon Steels

II.I Eurocode properties

The following sections describe the mechanical properties of carbon steel. The mechanical properties of stainless steel are given in Annex C of Eurocode 3. AISC steel design manual also lists Eurocode mechanical properties in the Appendix.

II.I.I Strength and deformation properties

The stress-strain relationship for carbon steel at elevated temperatures is showed in Figure II.1.

The following parameters define the shape of this characteristic:

$f_{y,\theta}$ effective yield strength;
$f_{p,\theta}$ proportional limit;
$E_{a,\theta}$ slope of the linear elastic range;
$\varepsilon_{p,\theta}$ strain at the proportional limit;
$\varepsilon_{y,\theta}$ yield strain;
$\varepsilon_{t,\theta}$ limiting strain for yield strength;
$\varepsilon_{u,\theta}$ ultimate strain.

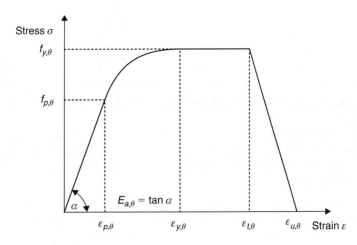

Fig. II.1 Stress-strain relationship for carbon steel at elevated temperatures

Table II.1 Formulas to determine the stress and the tangent modulus

Strain range	Stress σ	Tangent modulus
$\varepsilon \leq \varepsilon_{p,\theta}$	$\varepsilon E_{a,\theta}$	$E_{a,\theta}$
$\varepsilon_{p,\theta} < \varepsilon < \varepsilon_{y,\theta}$	$f_{p,\theta} - c + (b/a)[a^2 - (\varepsilon_{y,\theta} - \varepsilon)^2]^{0.5}$	$\dfrac{b(\varepsilon_{y,\theta} - \varepsilon)}{a[a^2 - (\varepsilon_{y,\theta} - \varepsilon)^2]^{0.5}}$
$\varepsilon_{y,\theta} \leq \varepsilon \leq \varepsilon_{t,\theta}$	$f_{y,\theta}$	0
$\varepsilon_{t,\theta} < \varepsilon < \varepsilon_{u,\theta}$	$f_{y,\theta}[1 - (\varepsilon - \varepsilon_{t,\theta})/(\varepsilon_{u,\theta} - \varepsilon_{t,\theta})]$	–
$\varepsilon = \varepsilon_{u,\theta}$	$0,00$	–
Parameters	$\varepsilon_{p,\theta} = f_{p,\theta}/E_{a,\theta} \quad \varepsilon_{y,\theta} = 0,02 \quad \varepsilon_{t,\theta} = 0,15 \quad \varepsilon_{u,\theta} = 0,20$	
Functions	$a^2 = (\varepsilon_{y,\theta} - \varepsilon_{p,\theta})(\varepsilon_{y,\theta} - \varepsilon_{p,\theta} + c/E_{a,\theta})$	
	$b^2 = c(\varepsilon_{y,\theta} - \varepsilon_{p,\theta})E_{a,\theta} + c^2$	
	$c = \dfrac{(f_{y,\theta} - f_{p,\theta})^2}{(\varepsilon_{y,\theta} - \varepsilon_{p,\theta})E_{a,\theta} - 2(f_{y,\theta} - f_{p,\theta})}$	

Table II.2 Reduction factors for stress-strain relationship of carbon steel at elevated temperatures

Steel temperature θ_a	Reduction factor (relative to f_y) for effective yield strength $k_{y,\theta} = f_{y,\theta}/f_y$	Reduction factor (relative to f_y) for proportional limit $k_{p,\theta} = f_{p,\theta}/f_y$	Reduction factor (relative to f_y) for design yield strength $k_{p0,2,\theta} = f_{p0,2,\theta}/f_y$	Reduction factor (relative to E_a) for the slope of the linear elastic range $k_{E,\theta} = E_{a,\theta}/E_a$
20°C	1,000	1,000	1,000	1,000
100°C	1,000	1,000	1.000	1,000
200°C	1,000	0,807	0.890	0,900
300°C	1,000	0,613	0.780	0,800
400°C	1,000	0,420	0.650	0,700
500°C	0,780	0,360	0.530	0,600
600°C	0,470	0,180	0.300	0,310
700°C	0,230	0,075	0.130	0,130
800°C	0,110	0,050	0.070	0,090
900°C	0,060	0,0375	0.050	0,0675
1000°C	0,040	0,0250	0.030	0,0450
1100°C	0,020	0,0125	0.020	0,0225
1200°C	0,000	0,0000	0.000	0,0000

Formulas to determine the stress and the tangent modulus at a given strain are given in Table II-1.

Table II-2 gives the reduction factors for the stress-strain relationship for steel at elevated temperatures given in Figure II.1. Linear interpolation may be used for values of the steel temperature intermediate to those given in the table. The reduction factors are defined as follows:

- effective yield strength, relative to yield strength at 20°C: $k_{y,\theta} = f_{y,\theta}/f_y$
- proportional limit, relative to yield strength at 20°C: $k_{p,\theta} = f_{p,\theta}/f_y$
- design yield strength, relative to yield strength at 20°C: $k_{p0,2,\theta} = f_{p0,2,\theta}/f_y$
- slope of linear elastic range, relative to slope at 20°C: $k_{E,\theta} = E_{a,\theta}/E_a$

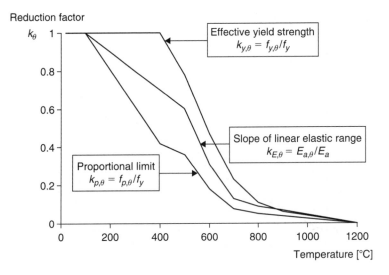

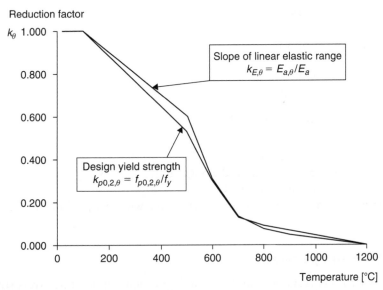

Fig. II.2 Reduction factors for the stress-strain relationship of carbon steel at elevated temperatures

The variation of these reduction factors with temperature is illustrated in Figure II.2.

II.1.2 *Thermal elongation*

The relative thermal elongation of steel should be determined from the following:

– for $20°C \leq \theta_a < 750°C$:

$$\Delta l/l = 1{,}2 \times 10^{-5}\theta_a + 0{,}4 \times 10^{-8}\theta_a^2 - 2{,}416 \times 10^{-4}$$

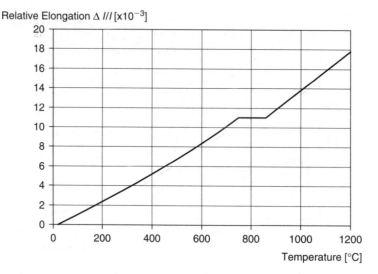

Fig. II.3   Relative thermal elongation of carbon steel as a function of the temperature

– for $750°C \leq \theta_a \leq 860°C$:

$$\Delta l/l = 1,1 \times 10^{-2}$$

– for $860°C < \theta_a \leq 1200°C$:

$$\Delta l/l = 2 \times 10^{-5}\theta_a - 6,2 \times 10^{-3}$$

where: l is the length at 20°C;
$\quad\quad\Delta l$ is the temperature induced elongation;
$\quad\quad\theta_a$ is the steel temperature [°C].

The variation of the relative thermal elongation with temperature is illustrated in Figure II.3.

II.2   ASCE properties

ASCE lists two sets (version 1 and version 2) of high temperature stress-strain relationships for structural steel. The first set (version 1) more suited for cases where load bearing is a primary function and is applicable for reinforcing steel or concrete-filled steel. The second set is recommended for structural steel. (More conservative than Version 2, but may be used for reinforcing steel or concrete-filled steel, where the role of steel in carrying the load at failure point is secondary).

II.2.1   Stress strain relations for steel (Version 1)

for $\varepsilon_S \leq \varepsilon_P$

$$F_y = \frac{f(T, 0.001)}{0.001}\varepsilon_s$$

where

$$\varepsilon_p = 4 \times 10^{-6} F_{y0}$$

and

$$f(T, 0.001) = (50 - 0.04T) \times \{1\exp[(30 + 0.03T)\sqrt{(0.001)}]\} \times 6.9$$

for $\varepsilon_S > \varepsilon_P$

$$F_y = \frac{f(T, 0.001)}{0.001} \varepsilon_p + f[T, (\varepsilon_s - \varepsilon_p + 0.001)] - f(T, 0.001)$$

where

$$f[T, (\varepsilon_s - \varepsilon_p + 0.001)] = (50 - 0.04T)$$

$$\times \{1\exp[(-30 + 0.03T)\sqrt{(\varepsilon_s - \varepsilon_p + 0.001)}]\} \times 6.9$$

and ε_s = strain in steel
ε_p = strain pertaining to proportional stress-strain relation

II.2.2 Stress strain relations for steel (Version 2)

(Less conservation then Version 1, recommended for structural steel, in particular where the rope of the steel in carrying the load is primary.)

for $\varepsilon_S \leq \varepsilon_P$

$$f_T = E_T \varepsilon_s$$

where

$$\varepsilon_p = \frac{0.975 f_{yT} - 12.5(f_{yT})^2}{E_T - 12.5 f_{yT}}$$

for $\varepsilon_S > \varepsilon_P$

$$f_T = (12.5\varepsilon_S + 0.975) f_{yt} - \frac{12.5(f_{yT})^2}{E_T}$$

In the above equation the yield strength f_{YT} is given by the following equations:

for $0 < T \leq 600°C$

$$f_{yT} = \left[1.0 + \frac{T}{900\ln\left(\dfrac{T}{1750}\right)} \right] f_{y0}$$

for $600°C < T \le 1000°C$

$$f_{yT} = \frac{340 - 0.34T}{T - 240} f_{y0}$$

and the modulus of elasticity is:
for $0 < T \le 600°C$

$$E_T = \left[1.0 + \frac{T}{2000\ln\left(\dfrac{T}{1100}\right)} \right] E_0$$

for $600°C < T \le 1000°C$

$$E_T = \frac{690 - 0.69T}{T - 53.5} E_0$$

where ε_s = strain in steel
ε_p = strain pertaining to proportional stress-strain relation
f_{y0} = yield strength of steel at room temperature
f_{yT} = yield strength of steel at temperature T
E_0 = modulus of elasticity at room temperature
E_T = modulus of elasticity at temperature T

II.2.3 Coefficient of thermal expansion

for $T < 1000°C$

$$\alpha_s = (0.004T + 12) \times 10^{-6}°C^{-1}$$

for $T \ge 1000°C$

$$\alpha_s = 16 \times 10^{-6}°C^{-1}$$

where α_s = coefficient of thermal expansion
T = temperature

Bibliography

AISC (2005), *Specification for structural steel buildings*, American Institute of Steel Construction, Inc., Chicago, IL.

Al-Jabri, K S, Davison, J B & Burgess, I W, (2008), *Performance of beam-to-column joints in fire – A review*, Fire Safety Journal, 42, 50–62

Anderberg, Y (2002), *Structural behaviour and design of partially fire-exposed slender steel columns*, Proc. 2nd int. Workshop "Structures in Fire", Univ. of Canterbury, Christchurch, P. J. Moss ed., 319–336

ASCE (2005), *Minimum design loads for buildings and other structures (ASCE Standard 7-05)*, American Society of Civil Engineers, Reston, VA.

ASCE (2005), *Standard calculation methods for structural fire protection (ASCE/SFPE Standard 29-05)*, American Society of Civil Engineers, Reston, VA.

ASTM (2005), *Standard methods of fire tests of building construction and materials (ASTM Standard E119-05)*, American Society for Testing and Materials, West Conshohocken, PA.

ASTM (2005), *Standard methods of fire tests for determining effects of large hydrocarbon pool fires on structural members and assemblies, (ASTM Standard E1529-05)*, American Society for Testing and Materials, West Conshohocken, PA.

Buchanan, A. H. (2001), *Structural design for fire safety*, John Wiley and Sons, Ltd., Chichester, UK.

Burgess, I W, El Rimawi, J & Plank R G (1991), *Studies of the Behaviour of Steel Beams in Fire*, J. Construct. Steel Research, 19, 285–312

Cadorin, J-F & Franssen, J-M (2003), *A tool to design steel elements submitted to compartment fires – OZone V2. Part 1: pre- and post-flashover compartment fire model*, Fire Safety Journal, Elsevier, 38, 395–427

Cadorin, J-F, Pintea, D, Dotreppe, J-C & Franssen, J-M (2003), *A tool to design steel elements submitted to compartment fires – OZone V2. Part 2: Methodology and application*, Fire Safety Journal, Elsevier, 38, 439–451.

CEB (1991), *Fire Design of Concrete Structures in accordance with CEB/FIP Model Code 90 (Final Draft)*, C.E.B., Paris, Bulletin d'Information n° 208, pp 188.

ECCS (1983), *European Recommendations for the Fire Safety of Steel Structures*, ECCS – Technical Committee 3 – Fire Safey of Steel Structures, Elsevier, Amsterdam, pp 106.

ECCS (1995), ECCS Technical Committee 3, *Fire Resistance of Steel Structures*, ECCS Publication No. 89, European Convention for Constructional Steelwork, Brussels, Belgium.

ECCS (2001), *Model Code on Fire Engineering*, ECCS – Technical Committee 3 – Fire Safey of Steel Structures, European Convention for Constructional Steelwork, First Edition, N° 111, May 2001, Brussels, Belgium.

Ellingwood, B R (2005), *Load combination requirements for fire-resistant structural design*, J. Fire Protection Engrg. 15(1):43–61.

El-Rimawi, J A, Burgess, I W & Plank, R J (1996), *The Treatment of Strain Reversal in Structural Members During the Cooling Phase of a Fire*, J. Construct. Steel Res., Vol. 37, No. 2, 115–135

ENV 13381-1 (2002), *Test methods for determining the contribution to the fire resistance of structural members – Part 1: Horizontal protective membranes*, European Committee for Standardization, Brussels

ENV 13381-2 (2002), *Test methods for determining the contribution to the fire resistance of structural members – Part 2: Vertical protective membranes*, European Committee for Standardization, Brussels

ENV 13381-4 (2002), *Test methods for determining the contribution to the fire resistance of structural members – Part 4: Applied protection to steel structural elements*, European Committee for Standardization, Brussels

Eurocode (2005), *Eurocode 3: Design of Steel Structures, Part 1.2: General Rules, Structural fire design*, European Committee for Standardisation, Brussels, Belgium.

Eurocode (2005), *Eurocode 4: Design of Composite Steel and Concrete Structures, Part 1-2: General Rules, Structural Fire Design*, European Committee for Standardisation, Brussels, Belgium.

FEMA (2002), *World Trade Center Building Performance Study: Data Collection, Preliminary Observations, and Recommendations*, Federal Emergency Management Agency (FEMA), Federal Insurance and Mitigation Administration, Washington, DC.

Franssen, J-M (1987), *Etude du Comportement au Feu des Structures Mixtes Acier-Béton*, Ph.D. thesis, Univ. of Liege, Collections de la F.S.A., N° 111, pp 276.

Franssen, J-M (1990), *The unloading of building materials submitted to fire*, Fire Safety Journal, Vol. 16, N° 3, 213–227

Franssen, J-M, Schleich, J-B & Cajot, L (1995), *A Simple Model for the Fire Resistance of Axially Loaded Members According to Eurocode 3*, J Construct. Steel Research, Vol. 35, 49–69

Franssen, J-M (1997), *Contributions à la Modélisation des Incendies dans les Bâtiments et de leurs effets sur les Structures*, Thèse d'agr. de l'ens. sup., Univ. of Liege, pp 391.

Franssen, J-M, Cajot, J-G & Schleich, J-B (1998), *Effects caused on the structure by localised fires in large compartments*, Proc. of the EUROFIRE conference, Brussels, pp, 19 in a CD-ROM,

Franssen, J-M, Talamona, D, Kruppa, J & Cajot, L-G (1998), *Stability of Steel Columns in Case of Fire : Experimental evaluation*, J Struct. Engng, ASCE, Vol. 124, No 2, 158–163

Franssen, J-M (2000), *Improvement of the Parametric Fire of Eurocode 1 based on Experimental Tests Results*, Proc. 6th int. Symp. on Fire Safety Science, IAFSS, Curtat, M. ed., Poitiers, 927–938

Franssen, J-M & Brauwers, L (2002), *Numerical Determination of 3D temperature fields in steel joints*, Proc. 2nd int. Workshop "Structures in Fire", Univ. of Canterbury, Christchurch, P. J. Moss ed., 1–20

Franssen, J-M (2003), *SAFIR. A Thermal/Structural Program Modelling Structures under Fire*, Proc. NASCC 2003, Baltimore, A.I.S.C. Inc.,

Garlock, M & Quiel, S E (2007), *The Behavior of Steel Perimeter Columns in a High-Rise Building under Fire*, Engineering Journal, AISC, 44(4).

Gulvanessian, H, Calgaro, J-A & Holicky, M (2002), *Designer's Guide to EN 1990. Eurocode: Basis of structural design*, Thomas Telford Publishing, London

Hasemi, Y & Tokunaga, T (1984), *Flame Geometry Effects on the Buoyant Plumes from Turbulent Diffusion Flames*, Fire Science and Technology, Vol. 4, N° 1.

Hasemi, Y & Tokunaga, T, Fire Science and Technology, 4, 15.

Hasemi, Y, Yokobayashi, Y, Wakamatsu, T & Ptchelintsev, A (1995), *Fire Safety of Building Components Exposed to a Localized Fire* – Scope and Experiments on Ceiling/Beam System Exposed to a Localized Fire, ASIAFLAM's 95, Hong Kong

Heskestad, G (1983), Fire Safety Journal, 5, 103.

Heskestad, G (1995), *Fire Plumes*, The SFPE Handbook of Fire Protection Engineering, 2nd Edition, SFPE-NFPA.

IBC (2006), *International Building Code, 2006 Edition*, International Code Council, Country Club Hills, IL, 2006.

ICC (2003), *ICC Performance Code for Buildings and Facilities*, International Code Council, Country Club Hills, IL.

ISO (2002), *ISO Standard 834: Fire resistance tests – Elements of building construction*, International Organization for Standardization, Geneva.

Joyeux, D. & Zhao, B. (1999), *Analyse du comportement de la structure porteuse du bâtiment de stockage automatise Procter & Gamble*, Centre technique Industriel de la Construction Métallique, CTICM – France

Kamikawa, D, Hasemi, Y, Wakamatsu, T & Kagiva, Y (2001), *Experimental flame heat transfer and surface temperature correlations for a steel column exposed to a localized fire*, Ninth Interflam conf., Interscience Ltd

Kirby, B R, Lapwood, D G & Thomson G (1986), *The reinstatement of Fire damaged Steel and Iron Framed Structures*, British Steel Corporation Swinden Laboratories.

Kirby, B R (1995), *The Behaviour of High-Strength Grade 8.8 Bolts in Fire*, J. Construct. Steel Research, 33, 3–38

Kodur, V K R, and Harmathy, T. Z. (2002), *Properties of Building Materials*, Chapter 10 of "*The SFPE Handbook of Fire Protection Engineering*", Third Edition, Society of Fire Protection Engineers, Bethesda, MA.

Latham, D J & Kirby, B R (1990), *Elevated Temperature Behaviour of Welded Joints in Structural Steel Works*, ECCS Research Project 7210-SA/824(F6.3/90)

Lie, T T (1992) (Editor), *Structural Fire Protection. ASCE Manuals and Reports of Engineering Practice*, No 78. American Society of Civil Engineers, New York.

Lie, T T (2002) *Fire Temperature-Time Relations*. Chapter 4-8 of "*The SFPE Handbook of Fire Protection Engineering*". Third Edition. Society of Fire Protection Engineers, USA.

Lim, L (2003), *Membrane Action in Fire Exposed Concrete Floor Systems*, Fire Engineering Research Report 03/2, Univ. of Canterbury, New Zealand

Lopes, N, Simões da Silva, L, Vila Real, P & Piloto, P (2004), *Proposals for the Design of Steel Beam-Columns under Fire Conditions, Including a New Approach for the Lateral-Torsional Buckling of Beams*, Computer & Structures, ELSEVIER, 82/17-19, 1463–1472

NBCC (2005), *National Building Code of Canada*, National Research Council of Canada. Ottawa, Canada.

NFPA (2003), *Building Construction and Safety Code, NFPA 5000*. National Fire Protection Association, Quincy, MA.

NFPA (2006), *Standard Methods of Tests of Fire Endurance of Building Construction and Materials*. NFPA 251. National Fire Protection Association, Quincy, MA.

PrEN 1991-1-2 (1992), *Eurocode 1 – Actions on Structures. Part 1-2 : General Actions – Actions on structures exposed to fire*, Final Draft Stage 49, European Committee for Standardization, Brussels, 10 January 2002.

PrEN 1993-1-2 (1993), *Eurocode 3 : Design of steel structures - Part 1.2 : General rules - Structural fire design*, European Committee for Standardization, Brussels, December 2003.

ProfilARBED (2001), *Background document on Parametric temperature-time curves according to Annex A of prEN1991-1-2 (24-08-2001)*, Profil ARBED document n° EC1-1-2/73, CEN/TC250/SC1/N298A, 22 October 2001.

Ptchelintsev, A, Hasemi, Y & Nikolaenko, M (1995), *Numerical Analysis of Structures exposed to localized Fire*, ASIAFLAM's 95, Hong Kong

Ranby, A (1998), *Structural Fire Design of Thin Walled Steel Sections*, J. Construct. Steel Research, 46, 303–4

Renaud, C (2003), *Modélisation numérique, expérimentation et dimensionnement pratique des poteaux mixtes avec profil creux exposés à l'incendie*, Ph. D. Thesis, INSA Rennes, France, pp. 334

Ruddy, J, Marlo, J P, Ioannides, S A & Alfawakhiri, F (2003), *Fire Resistance of Structural Steel Framing*, AISC Steel Design Guide 19, American Institute of Steel Construction, Chicago, IL, December 2003.

SAA (1990), *Fire Resistance Tests of Elements of Structure*. AS 1530.4-1990. Standards Association of Australia.

Schleich, J-B, Cajot, L-G, Kruppa, J, Talamona, D, Azpiazu, W, Unanua, J, Twilt, L, Fellinger, J, Van Foeken, R-J & Franssen, J-M (1998), *Buckling curves of hot rolled H steel sections submitted to fire*, C.E.C., EUR 18380 EN, pp. 333

Schleich, J-B, Cajot, L-G, Pierre, M, Brasseur, M, Franssen, J-M, Kruppa, J, Joyeux, D, Twilt, L, Van Oerle, J & Aurtenetxe, G (1999), *Development of design rules for steel structures subjected to natural fires in large compartments*, European Commission, technical steel research, Report EUR 18868 EN, pp 207.

SFPE (1995), *A Practical User's Guide to Fires-T3, A Three-Dimensional Heat-Transfer Model Applicable to Analyzing Heat Transfer through Fire Barriers and Structural Elements*, Society of Fire Protection Engineers, Task Group on Documentation of Fire Models, Bethesda, MD.

SFPE (2000), *Engineering Guide to Performance-Based Fire Protection*. Society of Fire Protection Engineers, Bethesda, Maryland.

SFPE (2002), *Guide to performance-based fire protection analysis and design of buildings*, Society of Fire Protection Engineers, Bethesda, Maryland.

SFPE (2004), *SFPE Handbook of Fire Protection Engineering*. Society of Fire Protection Engineers, SFPE.

SFPE (2004), Engineering *Guide: fire exposures to structural elements*, Society of Fire Protection Engineers, Washington, DC.

Talamona D (1995), *Flambement De Poteaux Métalliques Sous Charges Excentrées, A Haute Température*, Ph. D. thesis, Univ. Blaise Pascal – Ecole doctorale sciences pour l'ingénieur de Clermont-Ferrand, N° D.U. 726, EDSPIC: 85.

Talamona, D, Kruppa, J, Franssen, J-M & Recho, N (1996), *Factors influencing the behavior of steel columns exposed to fire*, J Fire Protection Engng, 8(1), 31–43

Talamona, D, Franssen, J-M, Schleich, J-B & Kruppa, J (1997), *Stability of Steel Columns in Case of Fire: Numerical Modelling*, J. Struct. Engng, ASCE, Vol. 123, No. 6, 713–720

UL (2003), *Fire Tests of Building Construction and Materials*, UL 263. Underwriters Laboratories Inc, Northbrook, Illinois.

UL (2004), *Fire Resistance Directory*, Volume 1, Underwriters Laboratories Inc., Northbrook, IL, 2004.

ULC (2004), *Standard Methods of Fire Endurance Tests of Building Construction and Materials*. CAN/ULC-S101-04. Underwriters Laboratories of Canada, Toronto, Ontario, Canada.

Vila Real, P M M, Lopes, N, Simões da Silva, L, Piloto, P & Franssen, J-M (2003), *Towards a consistent safety format of steel beam-columns: application of the new interaction formulae at ambient temperature to elevated temperatures*, Steel & Composite Structures, an International Journal, Techno-Press, Vol. 3, No. 6, 383–401.

Vila Real, P M M; Lopes, N, Simões da Silva, L & Franssen, J-M (2004), *Lateral-Torsional Buckling of Unrestrained Steel Beams Under Fire Conditions: Improvement of EC3 Proposal*, Computer & Structures, ELSEVIER, 82/20-21, 1737–1744

Vila Real, P M M , Simoes da Silva, L, Lopes, N & Franssen J-M (2005), *Fire resistance of unrestrained welded steel beams submitted to lateral-torsional buckling*, EUROSTEEL 2005, 4th European Conference on Steel and Composite Struct., Hoffmeister B & Hechler O ed., Aachen, 119–126

Wainman, D E & Kirby, B R (1988), *Compendium of UK Standard Fire Test Data. Unprotected Structural Steel – 1*, BRE, BSC Swinden Laboratories.

Wakamatsu, T, Hasemi, Y, Yokobayashi, Y & Ptchelintsev, A (1996), *Experimental Study on the Heating Mechanism of a Steel Beam under Ceiling exposed to a localized Fire*, Proc. 7th Interflam Conf., Cambridge, 509–518.

Wakamatsu, T, Hasemi, Y, Kagiya, K & Kamikawa, D (2002), *Heating Mechanism of Unprotected Steel Beam Installed beneath Ceiling and Exposed to a Localized Fire: Verification using the real-scale experiment and effects of the smoke layer*, Proc. 7th IAFSS Symp., Worcester Polytecnic Inst., Worcester, MA, 1099–1110.

Wang, Z (2004), *Heat Transfer Analysis of Insulated Steel Members Exposed to Fire*, Masters thesis, School of Civil and Env. Engng, NTU, Singapore.

Welch, S, Miles, S, Kumar, S, Lemaire, T & Chan, A (2008), FIRESTRUC – Integrating advanced three-dimensional modelling methodologies for predicting thermo-mechanical behaviour of steel and composite structures subjected to natural fires, Proc. 9th IAFSS symposium, Karlsruhe.

Wickström, U, *Application of the standard fire curve for expressing natural fires for design purposes*, Science and Engineering, ASTM, STP 882.

Wickström, U (1985), *Temperature analysis of heavily-insulated steel structures exposed to fire*, Fire Safety Journal, Vol. 5, 281–285.

Wickström, U (2001), *Calculation of heat transfer to structures exposed to fire – Shadow effects*, Ninth Interflam conf., Interscience Ltd, 451–460.

Zaharia R & Franssen J-M (2002), *Fire design study case of a high – rise steel storage building*. Third European Conference on Steel Structures, EUROSTEEL 2002, 19–20 September, 2002, Coimbra, Portugal

Subject index